Collins

Cambridge Lower Secondary

Science

STAGE 8: WORKBOOK

Beverly Rickwood, Heidi Foxford,
Dorothy Warren, Aidan Gill

Collins

William Collins' dream of knowledge for all began with the publication of his first book in 1819.

A self-educated mill worker, he not only enriched millions of lives, but also founded a flourishing publishing house. Today, staying true to this spirit, Collins books are packed with inspiration, innovation and practical expertise. They place you at the centre of a world of possibility and give you exactly what you need to explore it.

Collins. Freedom to teach.

Published by Collins
An imprint of HarperCollins*Publishers*
The News Building
1 London Bridge Street
London
SE1 9GF

HarperCollins*Publishers*
Macken House,
39/40 Mayor Street Upper,
Dublin 1, DO1 C9W8,
Ireland

Browse the complete Collins catalogue at
www.collins.co.uk

10 9 8 7 6 5

ISBN 978-0-00-836432-8

MIX
Paper | Supporting responsible forestry
FSC™ C007454

This book contains FSC™ certified paper and other controlled sources to ensure responsible forest management.

For more information visit:
www.harpercollins.co.uk/green

British Library Cataloguing-in-Publication Data

A catalogue record for this publication is available from the British Library.

Cambridge International copyright material in this publication is reproduced under licence and remains the intellectual property of Cambridge Assessment International Education.

End-of-chapter questions and sample answers have been written by the authors. These may not fully reflect the approach of Cambridge Assessment International Education.

Authors: Beverly Rickwood, Heidi Foxford, Dorothy Warren, Aidan Gill
Development editors: Anna Clark, Lynette Woodward, Tony Wayte, Sarah Binns
Product manager: Joanna Ramsay
Content editor: Tina Pietron
Project manager: Amanda Harman
Copyeditors: Debbie Oliver, Naomi Mackay
Proofreader: Life Lines Editorial Services
Illustrator: Jouve India Private Limited
Cover designer: Gordon MacGilp
Cover artwork: Maria Herbert-Liew
Internal designer: Jouve India Private Limited
Typesetter: Jouve India Private Limited
Production controller: Lyndsey Rogers
Printed and bound by: Martins the Printers

The publishers gratefully acknowledge the permission granted to reproduce the copyright material in this book. Every effort has been made to trace copyright holders and to obtain their permission for the use of copyright material. The publishers will gladly receive any information enabling them to rectify any error or omission at the first opportunity.

Acknowledgements

(t = top, c = centre, b = bottom, r = right, l = left)

p 28tl Valery Evlakhov/Shutterstock, p 28tr Steve Byland/Shutterstock, p 28tc schankz/Shutterstock, p 28b Carlo 2020/Shutterstock, p 34t Andrea Izzotti/Shutterstock, p 34tc RLS Photo/Shutterstock, p 34bc Artem Rudik/Shutterstock, p 34b Choksawatdikorn/Shutterstock, p 40 Salvacampillo/Shutterstock, p 45 Nasky/Shutterstock, p 146 ShutterStockStudio/Shutterstock, p 155 Sergey Merkulov/Shutterstock, p 156 bogadeva1983/Shutterstock, p 174l Safar Aslanov/Shutterstock, p 174cl MarcelClemens/Shutterstock, p 174cr robert_s/Shutterstock, p 174r maeching chaiwongwatthana/Shutterstock.

Contents

How to use this book

The questions in this workbook give you an opportunity to continue to learn and test your knowledge and recall at home.

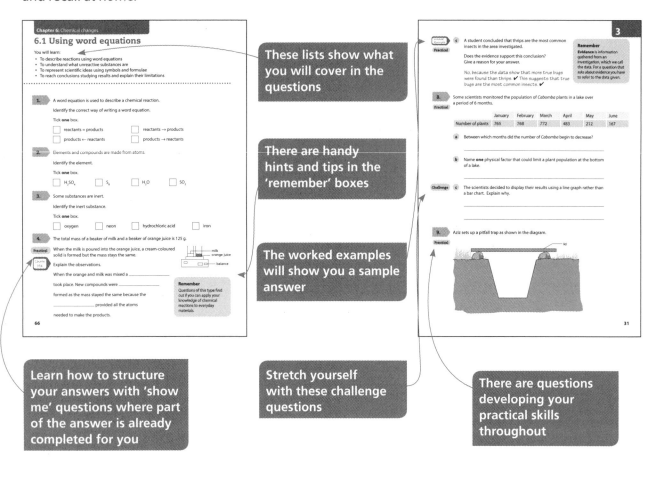

These lists show what you will cover in the questions

There are handy hints and tips in the 'remember' boxes

The worked examples will show you a sample answer

Learn how to structure your answers with 'show me' questions where part of the answer is already completed for you

Stretch yourself with these challenge questions

There are questions developing your practical skills throughout

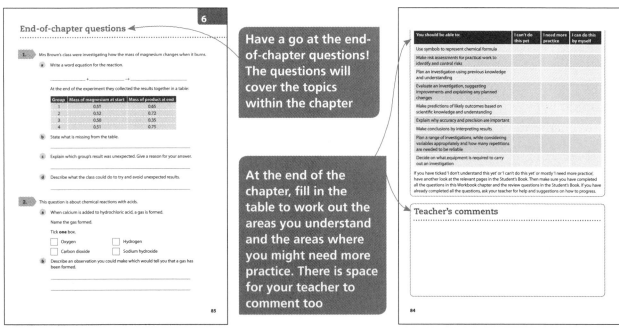

Have a go at the end-of-chapter questions! The questions will cover the topics within the chapter

At the end of the chapter, fill in the table to work out the areas you understand and the areas where you might need more practice. There is space for your teacher to comment too

Biology

Chapter 1: Respiration and moving

Chapter 2: Nutrition

Chapter 3: Ecosystems

1.1 Blood

You will learn:

- To describe and understand the role of the elements that make up blood
- To identify a hypothesis as testable
- To make predictions based on scientific knowledge and understanding

1. Complete the sentences using words from the list.

blood	cells	circulation	respiration	vessels	water

The circulatory system carries _____ around your body in tubes called

blood _____ . The movement of blood around your body is called

your _____ .

2. Why do cells in the body need oxygen?

_____

3. Draw **one** line to match the part of the blood with its function.

Part of blood

Plasma

Red blood cells

White blood cells

Function

Transport oxygen in the blood

Destroy microorganisms

Transports carbon dioxide and digested food

4. Amina takes her pulse. She counts 19 beats in 15 seconds.

Calculate her pulse rate in beats per minute (bpm).

First, calculate how many lots of 15 seconds are in 60 seconds: $60 \div 15 = 4$.

This means you must multiply the number of beats by 4 to get the number of beats per minute.

Therefore, $19 \times 4 = 76$ bpm. ✔

5. Amina thinks that pulse rate is affected by exercise. She believes that a person's pulse rate will increase as the intensity of exercise increases.

a What is Amina's hypothesis?

_____

b Write down Amina's prediction.

_____

_____

c Amina begins to plan a method to test her hypothesis. She decides to change the intensity of exercise by walking, jogging and sprinting.

Describe a method Amina could use to test her hypothesis.

_____

_____

d Amina did her experiment once and collected one set of data for each of walking, running and sprinting.

Describe one way Amina could make her results more reliable.

_____

_____

6. Describe two ways red blood cells are adapted to their function.

_____

_____

7. Anaemia is a condition when a person has lower levels of haemoglobin in their blood. Anaemia can cause people to feel short of breath and tired with lack of energy.

Challenge

Explain why anaemia can cause these symptoms.

_____

_____

8. Phil is investigating the effects of exercise on the body. His hypothesis is that running is better for a person's health than cycling.

a Decide whether Phil's hypothesis is testable.

b Explain your answer to **a**.

1.2 Human respiratory system

You will learn:

- To describe the structure and function of the human respiratory system
- To describe how oxygen and carbon dioxide circulate between the blood and the air in the lungs
- To know that aerobic respiration occurs in animals and plants and releases energy in a controlled way
- To be able to use the summary word equation for aerobic respiration
- To make predictions based on scientific knowledge and understanding
- To use results to describe the accuracy of predictions
- To present and interpret scientific enquiries correctly

● ●

1. The diagram shows the parts of the human respiratory system.

Name the parts labelled A, B and C.

A: _____

B: _____

C: _____

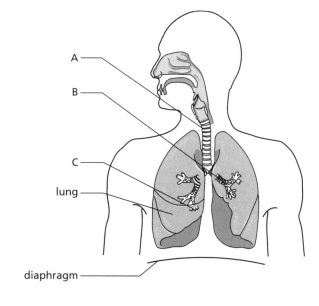

2. Complete the sentences using words from the list.

| outwards | inwards | contract | relax | decreases | increases |

When we inhale, intercostal muscles between the ribs _____ and move the

ribcage upwards and _____ . Muscles in the diaphragm contract and flatten it.

This _____ the volume of the chest.

3. Oliver is investigating the effects of exercise on breathing rate. He measures his breathing rate every minute while running for 5 minutes. He repeats this three times. Oliver predicts that his breathing rate will increase when he runs. The table shows his results.

Repeat 1	Repeat 2	Repeat 3
12	13	12
22	23	22
24	24	23
26	25	25
28	27	28
30	29	30

a What variable does he change? _____

b What variable does he measure? _____

c Do Oliver's results support his prediction?

☐ yes

☐ no

d Write a conclusion for Oliver's investigation.

_____

e Evaluate Oliver's results. In your answer you should comment on:

• the reliability of the results

• how to improve the results.

_____

_____

_____

4. Yuri makes a model of the lungs as shown below.

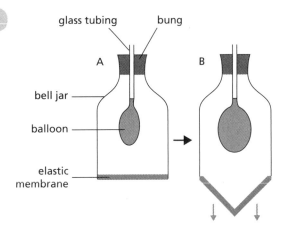

State what part of the human respiratory system is represented by:

a The elastic membrane _____

b The bell jar _____

c The balloon _____

d Explain at least one strength and one weakness of this model.

_____

_____

_____

_____

_____

_____

5. Explain how gas exchange happens in the alveoli.

There is more oxygen in the alveoli than in the capillaries so oxygen moves

from the _____ into the _____ .

There is more carbon dioxide in the capillaries than in the lungs so carbon dioxide

moves from the _____ into the _____ .

6. What is the name of the process where particles spread out from an area of higher concentration to an area of lower concentration?

_____

7. Describe how gas exchange happens in plants.

Worked Example Plant leaves have holes called stomata. When the stomata are open gases can move into and out of the leaf.

8. Describe what happens in the process of aerobic respiration.

_____

9. Where does aerobic respiration take place? Tick **one** box.

☐ Only in muscles ☐ Only in the leaves of plants

☐ Only in the lungs ☐ In the mitochondria of plant and animal cells

10. Complete the word equation for aerobic respiration.

_____ + oxygen ⟶ carbon dioxide + _____

Remember
Energy is not written as a product of respiration as it is not a substance.

11. Explain the difference between respiration and breathing.

_____

_____

_____

12. Long distance cyclists often drink liquids that contain glucose.
Explain how this helps them perform well.

Challenge

_____

_____

_____

1.3 Skeleton, joints and muscles

You will learn:

- To identify ball-and-socket and hinge joints
- To explain what makes bones move at hinge joints
- To make predictions based on scientific knowledge and understanding
- To use results to describe the accuracy of predictions

. . . .

1. List **three** functions of the skeleton.

1. _____

2. _____

3. _____

2. The diagram shows a human skeleton.

a Which two letters show hinge joints?

_____

Show Me

b What type of joint is **D?** Explain your answer.

The joint is a ball

_____ .

This joint allows the leg

_____ .

3. Describe the type of movement allowed by a hinge joint.

_____

4. **a** Explain how bones at a hinge joint are moved by an antagonistic pair of muscles.

_____

b Complete the table by writing 'contracts' or 'relaxes' to show the action of the muscles in the arm.

Muscle	Bending the arm	Straightening the arm
Biceps		
Triceps		

5.

Challenge

What happens when an athlete 'pulls a muscle'?

_____

_____

6.

Practical

Fatma is investigating how calcium affects the strength of bones. Calcium is a mineral found in bones.

Fatma uses a chemical to remove some of the calcium from four bones.

She then tries to bend each bone to find out if it breaks easily.

a What question is Fatma trying to answer?

_____

Fatma does some research and finds that calcium is needed for strong bones.

b Write a prediction for Fatma's investigation.

_____

c Fatma finds that each of the four bones breaks very easily after some of the calcium has been removed. Does this evidence support her prediction?

_____

Self-assessment

Tick the column which best describes what you know and what you are able to do.

What you should know:	I don't understand this yet	I need more practice	I understand this
The human circulatory system contains the heart, blood and blood vessels			
The circulatory system ensures that all cells have enough oxygen and food, and removes waste products			
Blood plasma carries dissolved food substances, and waste products (such as carbon dioxide)			
Red blood cells are adapted to carry oxygen			
White blood cells destroy microorganisms			
Breathing is the movement of the diaphragm and intercostal muscles, and it causes the lungs to expand and shrink			
The walls of the alveoli and blood vessels in the lungs are only one cell thick, to allow gas exchange to happen quickly by diffusion			
Diffusion is the overall movement of a certain type of particle from where there are many of them to where there are fewer			
Aerobic respiration occurs in mitochondria and requires glucose and oxygen, and produces carbon dioxide and water			
Gas exchange in plants happens in tiny holes in their leaves called stomata			
Each stoma can be opened or closed			
When the stomata are open, gases (such as oxygen, carbon dioxide and water vapour) can move into and out of the leaf			
Aerobic respiration can be shown using a word equation: oxygen + glucose → carbon dioxide + water			

	I can't do this yet	I need more practice	I can do this by myself
Ciliated epithelial cells help keep the lungs clean			
There are different types of joints in the skeleton and these allow different types of movement			
Antagonistic pairs of muscles move bones at a hinge joint			
You should be able to:	**I can't do this yet**	**I need more practice**	**I can do this by myself**
Identify whether a given hypothesis is testable			
Make a prediction			
Compare the results with a prediction in order to reach a conclusion			
Evaluate results in terms of reliability			
Explain observations using scientific language			

If you have ticked 'I don't understand this yet' or 'I can't do this yet' or mostly 'I need more practice', have another look at the relevant pages in the Student's Book. Then make sure you have completed all the questions in this Workbook chapter and the review questions in the Student's Book. If you have already completed all the questions, ask your teacher for help and suggestions on how to progress.

Teacher's comments

· ·

End-of-chapter questions

1. The diagram on the right shows the ribcage.

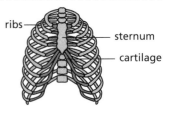

a Name an organ protected by the ribcage that pumps blood.

_____

b Red blood cells are found in the blood. Describe the function of red blood cells.

_____

c Cartilage joins each rib to the sternum. This allows the ribs to move.

Explain why it is important that the ribs can move.

_____

_____

d Describe how ciliated epithelial cells help keep the lungs clean.

_____

_____

_____

2. Some students measured their lung volume using the apparatus shown in the diagram.

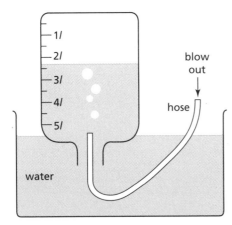

Each student was told to take a deep breath in and blow out into the tube.
Each student had their lung volume measured three times.

The results are shown in the table.

	Lung volume (litres)		
Student	Trial 1	Trial 2	Trial 3
A	3.5	3.2	3.6
B	3.7	3.9	4.0
C	3.1	3.4	3.0

a Evaluate the reliability of the data using the data in the table.

_____

_____

b A student makes a prediction that the taller a person, the greater their lung volume. Student A was 150 cm tall, student B was 154 cm and student C was 148 cm.

Does the data in the table support the student's prediction? Explain why.

_____

_____

3. The diagram below shows the muscles in the arm.

a Write in the names of muscles A and B in the diagram.

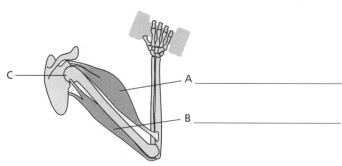

b Explain how the muscles work together antagonistically to bend and straighten the arm.

_____

_____

c Name the type of joint labelled C.

_____

d Describe the type of movement allowed by joint C.

_____

2.1 A balanced diet

You will learn:

- To identify what makes a balanced diet for humans
- To describe the functions of the nutrients needed in a balanced diet
- That carbohydrates and fats can be used as an energy store in animals
- To identify and control risks in practical work
- To plan investigations, including fair tests, while considering variables

1. Draw a line to match each food group with the reason it is needed.

Food group	Reason food group is needed
Protein	To store energy
Carbohydrate	For growth and repair
Fat	To keep our bodies healthy and working properly
Vitamins	To store energy

2. Name the main food group we get from:

a fish _____

b rice _____

3. List **two** ways the body uses energy from food.

1. _____   2. _____

4. Which of the following foods is a good source of fibre? Tick **one** box.

☐ Yoghurt   ☐ Eggs   ☐ Lentils   ☐ Chicken

5. Name a condition that results from too little fibre in the diet. _____

6.

Challenge

Water is a nutrient that we need to survive. Explain why.

_____

_____

_____

_____

Remember

Approximately 70% of the body is water. The chemical reactions in our cells take place in water.

7.

Practical

Osama is burning different types of nut to find out how much energy is released. He uses the apparatus shown below.

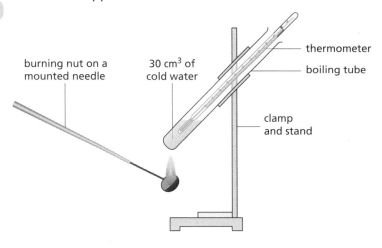

burning nut on a mounted needle

30 cm³ of cold water

thermometer

boiling tube

clamp and stand

Osama burns 2 grams of each type of nut. He measures the increase in the temperature of the water in the boiling tube.

Show Me

a Write down **two** safety precautions Osama needs to take to reduce the risks to himself.

Keeping clothes away from _____ .

Do not touch _____ .

Osama's results are shown in the table.

Type of nut	Temperature of water at start (°C)	Temperature of water at end (°C)	Rise in temperature (°C)
Pecan	16.3	26.8	10.5
Almond	17.2	21.5	
Cashew	17.7	23.7	

b Complete the table by calculating the change in the temperature for the almond and cashew nut.

c In Osama's investigation state:

The independent variable:

_____

The dependent variable:

_____

> **Remember**
> The independent variable is the variable that is changed and the dependent variable is the one that is measured or observed.

One variable that needs to be kept the same:

_____

_____

d Which type of nut contained the most energy? _____

8. Fatma is an Olympic athlete. Her friends suggest she should only eat foods very high in protein such as chicken, soya and eggs.

a Give **one** reason why an athlete needs extra protein in their diet.

_____

_____

b Explain why Fatma should **not** only eat foods very high in protein.

_____

_____

_____

_____

9. The table below shows the nutrition facts for a breakfast cereal for children.

	Per 100 g	Per 30 g serving
Energy	1647 kJ (389 kcal)	494 kJ (117 kcal)
Fat	2.8 g	0.8 g
Carbohydrates	84 g	25 g
of which sugars	32 g	9.6 g
Fibre	2.3 g	0.7 g
Protein	6.2 g	1.9 g
Salt	0.75 g	0.22 g

a Explain why food labels show amounts per 100 g.

_____

Challenge **b** Use evidence from the table to evaluate whether this cereal is healthy.

_____

_____

_____

10. The table shows how many kilojoules of energy are required for a 60 kg person to do different activities for **one** hour.

Activity	kJ
Standing in line	300
Playing cricket	1200
Playing football	1740
Running up stairs	3780

Use the information in the table to calculate:

a the number of kJ (kilojoules) needed for a 60 kg person to play cricket for **two** hours.

_____

b the number of kJ (kilojoules) needed for a 60 kg person to play football for **two** hours.

_____

c Explain why a football player might need to eat more carbohydrates than a cricket player.

_____

_____

_____

2.2 The effects of lifestyle on health

You will learn:

- To discuss the effects of lifestyle on humans' development and health

1. Obesity is a condition linked to what sort of diet? Tick **one** box.

☐ A diet high in fat and high-energy foods

☐ A diet low in fat

☐ A diet high in fibre

☐ A balanced diet

2. Which **two** of the following conditions are often caused by overeating fatty foods?

☐ Low blood pressure   ☐ Obesity

☐ High blood pressure   ☐ Scurvy

3. Give the main cause of obesity.

_____

4. Which of the following health problems are associated with obesity? Tick **two** boxes.

☐ low blood pressure   ☐ scurvy

☐ high blood pressure   ☐ type two diabetes

5. Explain how eating too much fat can cause heart disease.

Eating too much fat can cause blood vessels in the heart to become blocked ✓ which can cause heart disease ✓.

6. Ayesha is obese. Her doctor has recommended she should change her lifestyle by eating a healthy diet and exercising more.

Explain how exercise could help Ayesha become less obese.

_____

_____

7. Some people smoke tobacco. Name a substance found in tobacco smoke that

a is addictive: _____

b can cause cancer: _____

8. The graph below shows how an increase in the number of cigarettes smoked per day affects the number of deaths from lung cancer (per 100 000 men per year).

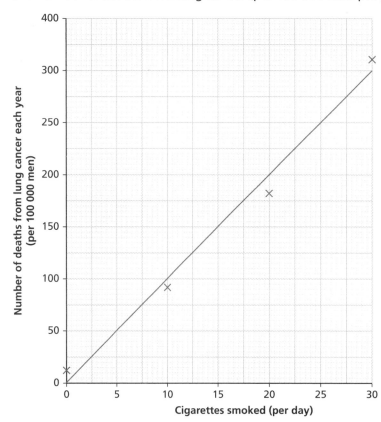

Show Me **a** Describe the trend shown by the graph.

The more cigarettes smoked per day,

the _____

_____

Challenge **b** Juan claims that smoking fewer than four cigarettes a day does not affect your health. Use the graph to give evidence that Juan's claim is **not** true.

_____

_____

9. Smoking damages alveoli. Explain how damage to alveoli causes poor gas exchange.

Challenge _____

_____

_____

Self-assessment

Tick the column which best describes what you know and what you are able to do.

What you should know:	I don't understand this yet	I need more practice	I understand this
A balanced diet contains the right amount of the different nutrients from a wide variety of different foods.			
You need fibre in your diet too			
Carbohydrates and fats store energy, which is measured in joules (J) or kilojoules (kJ)			
Eating too much carbohydrate or fat may cause weight gain and make people unhealthy			
For a balanced diet you need to eat foods from many sources			

What you should know:	I don't understand this yet	I need more practice	I understand this
Eating too much food that is rich in fat or carbohydrate can cause obesity			
Obesity can cause problems with the heart and blood vessels, so blood does not flow so well			
Tobacco smoke paralyses cilia, and contains an addictive drug called nicotine			
Tar in tobacco smoke causes cancer			

You should be able to:	I can't do this yet	I need more practice	I can do this by myself
I can identify independent, dependent and control variables			
I can use tables to present information			
I can look for trends and patterns in data			
I can interpret line graphs (and lines of best fit)			
I can write a risk assessment			

If you have ticked 'I don't understand this yet' or 'I can't do this yet' or mostly 'I need more practice', have another look at the relevant pages in the Student's Book. Then make sure you have completed all the questions in this Workbook chapter and the review questions in the Student's Book. If you have already completed all the questions, ask your teacher for help and suggestions on how to progress.

Teacher's comments

End-of-chapter questions

1. Malik is researching how carbohydrates are digested.

Challenge **a** State why we need carbohydrates in our diet.

_____

Challenge **b** Name two foods that contain carbohydrate.

_____

2. A person who is obese can have poor nutrition. Explain why.

_____

_____

3. Nasir has smoked tobacco for 10 years. He has developed a smoker's cough. Explain how smoking tobacco has caused Nasir to have a smoker's cough.

_____

_____

_____

3.1 Habitats and ecosystems

You will learn:

- To identify different ecosystems on Earth
- To recognise the range of habitats in an ecosystem
- To collect and record observations and measurements appropriately
- To sort organisms through testing and observation
- To plan investigations, including fair tests, while considering variables
- To choose experimental equipment and use it correctly
- To present and interpret scientific enquiries correctly

1. Draw **three** lines to match each organism to its habitat.

Organism	Habitat
Tilapia fish	Tree
Hawk	Soil
Worm	River

2. Describe **two** physical factors in a desert habitat.

1. _____

2. _____

3. Which statement best describes an organism's environment? Tick **one** box.

☐ The other organisms and physical factors in the surroundings

☐ The other animals in the surroundings

☐ The resources in the surroundings

☐ The amount of water in the surroundings

4.

Practical

Rajiv wants to find out which conditions maggots prefer to live in. He uses a large container that has four different areas. Each area has a different condition to the others, as shown in the diagram on the right.

The maggots can move easily between the different areas.

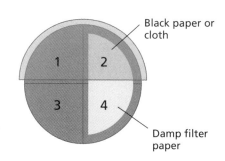

Black paper or cloth

Damp filter paper

Area 1: dark and dry
Area 2: dark and moist
Area 3: light and dry
Area 4: light and moist

Show Me **a** Write a plan that describes how Rajiv could find out which condition maggots prefer.

Place 20_____ in the centre of the container. Wait a set amount

of time and then measure _____ . Record the _____

and repeat three times. The area that most maggots go to will show

_____ .

b What type of graph or chart would be most suitable for displaying these results?

_____

5. Gabriella sees some tiny insects on a tree trunk.

Practical

a Name a piece of equipment that she could use to collect these insects.

_____

b Gabriella decides to make some detailed drawings of the insects she has collected.

What piece of equipment would help her observe the insects in detail?

_____

6. Draw **three** lines to match each piece of equipment to its correct use.

Equipment

| Explanations |

| quadrat | | jar buried in the ground to collect small organisms that walk on the ground |

| pooter | | square frame used to take samples of a population in a habitat |

| pitfall trap | | device used to carefully suck small organisms into a collecting jar |

A scientist investigated the frequency of different species of insect in an area of grassland. Twelve pitfall traps were used to collect 12 samples from the grassland.

The table shows the number of four types of insect found in each trap the next day.

Type of insect	Sample											
	1	2	3	4	5	6	7	8	9	10	11	12
Springtail				✓					✓			
Mantis	✓		✓			✓					✓	
True bug		✓✓		✓			✓	✓	✓	✓		✓
Thrip		✓			✓✓		✓	✓			✓	

a Use the results to complete the table to show the frequency of each type of insect found.

Type of insect	Frequency
Springtail	
Mantis	
True bug	
Thrip	

b Draw a bar chart to compare the frequency of each type of insect.

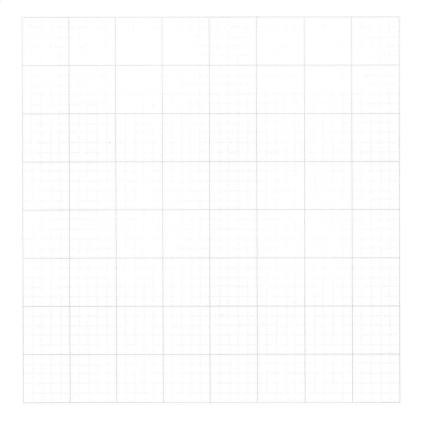

Worked Example

Practical

c A student concluded that thrips are the most common insects in the area investigated.

Does the evidence support this conclusion?
Give a reason for your answer.

No, because the data show that more true bugs were found than thrips. ✔ *This suggests that true bugs are the most common insects.* ✔

Remember

Evidence is information gathered from an investigation, which we call the data. For a question that asks about evidence you have to refer to the data given.

8.

Practical

Some scientists monitored the population of *Cabomba* plants in a lake over a period of 6 months.

	January	February	March	April	May	June
Number of plants	765	768	772	483	212	167

a Between which months did the number of *Cabomba* begin to decrease?

_____

b Name **one** physical factor that could limit a plant population at the bottom of a lake.

_____

Challenge

c The scientists decided to display their results using a line graph rather than a bar chart. Explain why.

_____

_____

9.

Practical

Aziz sets up a pitfall trap as shown in the diagram.

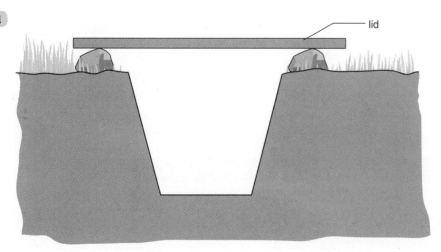

lid

a What sort of animals is a pitfall trap best for trapping?

_____

b What is the lid for?

_____

c Suggest a reason why pitfall traps should be checked regularly.

_____

3.2 Bioaccumulation in food chains

You will learn:

- To describe the effect of toxic substances in an ecosystem
- To use scientific understanding to evaluate issues
- To discuss the global environmental impact of science

1. Complete the sentences using the words from the box.

consumer	eats	energy	food	kills	producer	shelter

A food chain shows what _____ what in a habitat. It also shows

how _____ flows through organisms in a habitat. A food chain starts

with a _____ . These are organisms that can make their own

_____ using energy from the Sun.

2. The diagram below shows an aquatic food chain.

Phytoplankton   ⟶   Anchovy   ⟶   Barracuda

> **Remember**
> A producer is an organism that makes its own food. A consumer is an animal that eats other living things.

Use the information in the food chain to complete the table:

Predator:	Prey:	Producer:
_____	_____	_____

3. Which type of organisms can make their own food using energy from the Sun?

Tick **one** box only.

☐ Predators   ☐ Consumers

☐ Producers   ☐ Both consumers and producers

4. Describe the similarities and differences between herbivores, carnivores and omnivores.

Worked Example

Herbivores are animals that only eat plants.

Carnivores are those that only eat other animals.

Those that eat plants and animals are omnivores.

5. What do the arrows in a food chain represent?

_____

6. Lana says, 'humans could not survive without plants'. Is this true? Explain why.

Show Me

Yes, it is true, because plants form the start of the food chain and _____ on to consumers. We would _____ without plants.

7. Heston draws a food chain found in an Arabian desert.

Practical

Complete Heston's table by adding ticks to show what terms describe each organism along the food chain.

	Desert grass	→	Jerboa (rodent)	→	Cobra (snake)	→	Peregrine falcon (bird)
Producer							
Herbivore							
Carnivore							
Primary consumer							
Secondary consumer							
Top predator							
Prey							

8. Explain why there are fewer organisms at higher trophic levels.

_____

_____

9. A scientist is investigating the bioaccumulation of mercury in an ocean food chain. In this food chain, plankton is the producer and dolphin is the top predator.

The scientist measures the concentration of mercury in the tissues of each organism in the food chain. Their results are shown below.

Organism	Concentration of mercury in body tissue of organism (ppm)	
Dolphin	0.30 	
Squid	0.10 	
Shrimp	0.03 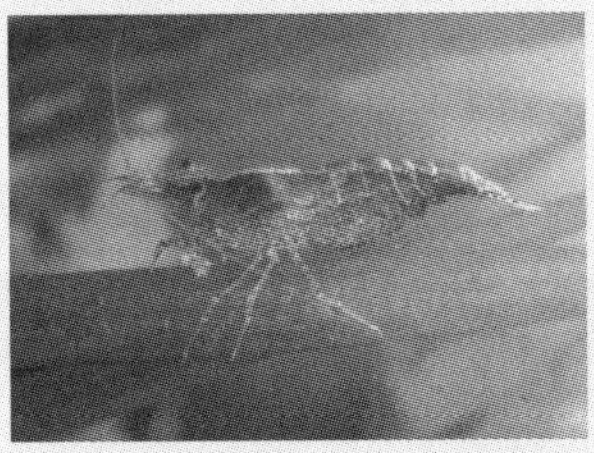	
Plankton	0.01 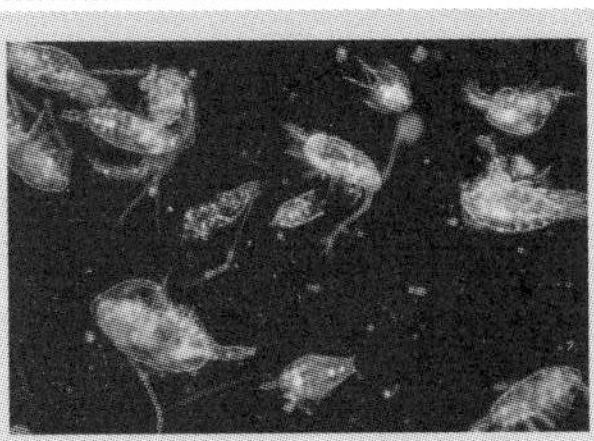	

a Explain what is meant by the term **bioaccumulation**.

_____

_____

Challenge **b** Suggest a reason why the dolphin was found to have the highest concentration of mercury in its body.

_____

_____

_____

3.3 Invasive species

You will learn:

- To describe the effects of new or invasive species on an ecosystem
- To describe the application of science in society, industry and research

1. The diagram below shows a food web in a pond.

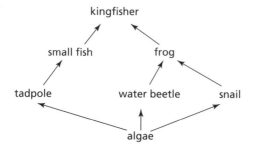

> **Remember**
> A primary consumer is the first consumer in a food chain and is always a herbivore.

a Use the food web to draw a food chain that has **four** trophic levels.

_____ → _____ → _____ → _____

b List all the organisms in the food web that compete for algae.

_____

c The number of algae in the pond increases. How is this likely to affect competition between the primary consumers? Tick **one** box only.

☐ competition will increase

☐ competition will decrease

☐ competition will stay the same

☐ it is impossible to tell

d Most of the algae in the pond die. Explain what will happen to the size of the population of snails.

The number of snails will _____ because there will be

_____ competition for the _____ and so the

snails will have less food and will be less likely to survive.

_____

2. Complete the table to show what resources organisms need to survive.
Add **one** tick to each row. The first row has been done for you.

Resource	Plants	Animals	Both plants and animals
Sunlight	✓		
Water			
A source of food			
Carbon dioxide			

3. Describe the difference between a **native** and a **non-native species**.

_____

4. Asian carp were introduced to North America from Eastern Russia as a food source for humans and for sport fishing. The numbers of Asian carp in North America increased rapidly after they were introduced. Asian carp grow and reproduce very quickly. They compete with native species for food and space and eat the eggs of native fish species.

a Suggest **two** reasons why Asian carp populations increased rapidly.

1. _____

2. _____

b Give **two** reasons why increases in the population of Asian carp could cause a decrease in native species of fish.

1. _____

2. _____

36

5.

Caulerpa is an invasive alga that grows in the sea. It is native to the Caribbean Sea and the Indian Ocean, but was accidently introduced into the Mediterranean Sea in waste water.

Caulerpa has now spread over large areas of seabed in the Mediterranean. Populations of native species in these areas have decreased or been wiped out.

Suggest how Caulerpa can cause a decrease in the number of species living in ecosystems.

_____

_____

Self-assessment

Tick the column which best describes what you know and what you are able to do.

What you should know:	I don't understand this yet	I need more practice	I understand this
An ecosystem is a set of living things that depend on one another and a set of physical environmental factors found in a particular area			
Examples of ecosystems include tropical rainforest, desert, grassland and coral reef			
A habitat is the place in an ecosystem where a certain organism can find the resources it needs			
An organism's environment is the physical factors and the other organisms in its surroundings			
A population is the number of organisms of one species living in a certain area			
Different sampling methods are used to find and count different types of organisms			
A trophic level is the stage at which an organism is found in a food chain			
The higher the trophic level an organism is in, the less energy is available			
Some poisonous substances do not break down quickly and so bioaccumulate in organisms			
Organisms compete for the resources they need			

You should be able to:	I can't do this yet	I need more practice	I can do this by myself
An invasive species is one that is new to an area and causes damage to the ecosystem			
If one organism is growing very well, it may use up the resources that other organisms need to survive			
Choose and use a variety of different sampling techniques to study organisms in their habitats			
Plan to collect reliable data			
Use data to make conclusions			
Understand energy flow and bioaccumulation			

If you have ticked 'I don't understand this yet' or 'I can't do this yet' or mostly 'I need more practice', have another look at the relevant pages in the Student's Book. Then make sure you have completed all the questions in this Workbook chapter and the review questions in the Student's Book. If you have already completed all the questions, ask your teacher for help and suggestions on how to progress.

Teacher's comments

· ·

End-of-chapter questions

1. The diagram shows a food chain.

Grass   →   Grasshopper   →   Mouse   →   Owl

a Name the producer in the food chain. _____

b Name a herbivore in the food chain. _____

c The number of mice in the food chain suddenly decreases.

Explain how this may affect the number of:

i) Grass plants

_____

_____

ii) Owls

_____

_____

2. A food chain shows 'what eats what'.

What do the arrows in a food chain represent?

_____

3. Juan says: 'All the energy in a food chain comes from the Sun.'

 Explain why this statement is true.

Energy from the Sun is stored in the biomass made by plants during photosynthesis. This energy is passed along the food chain when one organism consumes another.

Remember

'Biomass' is the mass of living material that organisms make for themselves. The biomass of an organism contains a store of energy.

4. Orangutans are apes that live in Borneo and Sumatra. An orangutan's preferred habitat is lowland forests.

a What is meant by the term habitat?

_____

b Give **two** examples of resources that orangutans get from their habitats.

_____

c Some of Borneo's forests have been affected by an invasive species, the spiked pepper tree. This species of tree is spreading at an estimated rate of 5-7 km per year, threatening native species.

i) Describe what is meant by the term 'invasive species'.

_____

ii) List two resources that the spiked pepper tree might compete for with native species.

_____

5. A farmer sprays a toxic chemical known as DDT onto a field next to a pond. A week later, DDT was detected in algae in the pond. A month later several large fish that had been living in the pond were found dead on the surface of the water. Scientists found high concentrations of DDT in the tissues of these large fish.

Explain how high concentrations of DDT ended up in the tissues of the large fish.

_____

_____

_____

_____

_____

Chemistry

4.1 Structure of an atom

You will learn:

- To describe the Rutherford model of the atom
- To know the charges on electrons, protons and neutrons
- To know how individual atoms are held together
- To discuss how scientific knowledge is developed over time

1. Atoms have no overall charge – they are neutral. Which one of the sentences below explains why?

Tick **one** box.

☐ They contain the same number of electrons and protons.

☐ They contain the same number of protons and neutrons.

☐ They contain neutrons, which have no charge.

☐ The nucleus contains the same number of positive and negative particles.

2. The electrostatic attraction, which holds atoms together, occurs between positive and negative charges. Which statement about positively charged and negatively charged particles is true?

Tick **one** box.

☐ Electrons are positively charged and protons are negatively charged.

☐ Electrons are negatively charged and neutrons are positively charged.

☐ Electrons are negatively charged and protons are positively charged.

☐ Neutrons are negatively charged and protons are positively charged.

3. This diagram models the inside of a carbon atom.

Label the diagram by choosing words from this list.

period	group	proton	neutron
	proton number		electron
		nucleus	

....

....

....

....

4. The following diagram shows the 'plum pudding' model of the atom. It was first suggested by the scientist J.J. Thomson in 1904.

a Complete these sentences about Thompson's model using words from this list.

negatively	neutrally	positively	electrons
protons	neutrons	nucleus	

The atom is a _____ charged ball with negatively charged

_____ in it. In 1912, the idea of having a _____

in the centre of the atom was not the accepted model.

b Suggest reasons why JJ Thomson used the plum pudding as an analogy for his model of the atom.

_____

_____

_____

_____

5. Complete the table to show the charges on each particle found in an atom:

Particle	Charge
Electron	
Neutron	
Proton	

6. Fill in the gaps in the sentences below describing Rutherford's model of the atom:

Show Me

In Rutherford's model of the atom the _____ and _____ are found inside the nucleus. The _____ are found orbiting the nucleus.

The atom is held together by the electrostatic force of _____ between the positive _____ and _____ electrons.

7. These diagrams show two different models of the atom.

Ernest Rutherford's model
of the atom (1911)

J.J. Thomson's plum
pudding model
of the atom (1904)

a State **one** similarity and **one** difference between the two models.

_____

_____

_____

b Describe how Rutherford proved that the plum pudding model was limited.

_____

_____

_____

_____

c The understanding of the model of the atom has developed through the collective understanding of scientists and the scrutiny of their theories over time. When a scientist puts forward a new model or scientific theory other scientists often try to repeat their work.

Suggest a reason why.

_____

_____

_____

4.2 Paper chromatography

You will learn:

- To describe the use of paper chromatography with substances
- To describe results in terms of any trends and patterns and identify any abnormal results
- To reach conclusions studying results and explain their limitations
- To evaluate experiments and investigations, and explain any improvements suggested

1. Draw a line to match the key word to its meaning. The first one has been done for you.

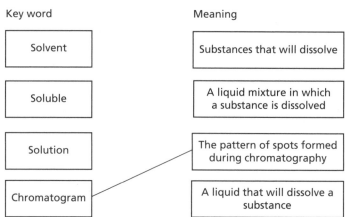

Key word | Meaning
- Solvent — Substances that will dissolve
- Soluble — A liquid mixture in which a substance is dissolved
- Solution — The pattern of spots formed during chromatography
- Chromatogram — A liquid that will dissolve a substance

2. Complete the sentences using words or phrases from this list.

| compounds | properties | mixtures | filtration |
| chromatography | melting points | dyes in ink | densities |

_____ is used to separate _____ of soluble

substances, often coloured substances like _____ or food colouring.

It works because the soluble substances have different _____.

3. The diagram shows the apparatus needed for paper chromatography.

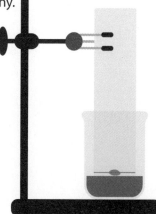

a Label the diagram by choosing words from the list.

| beaker | ink spot | solvent |
| chromatography paper | pencil line |

b Suggest a reason why the line is drawn in pencil and not pen.

_____

_____

4. A group of students investigated the food colouring in a range of sweets. The diagram shows the resulting chromatogram.

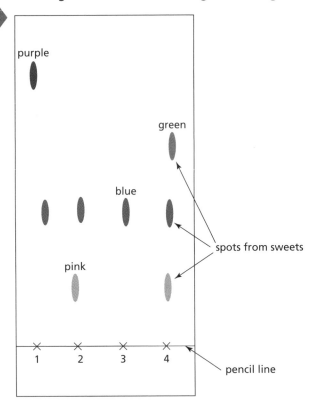

a Write down the number of substances found in sweet 4.

3 ✓

b The students think that one substance is found in all of the sweets.

Explain why.

They all show a blue spot in the same place ✓

c Explain how chromatography works.

As the solvent travels up the chromatography paper it takes the dissolved dyes with it ✓ because they are attracted to both the solvent and the paper ✓

The dyes travel different distances up the paper so they separate out. ✓ This is because there are differences in how much each dye is attracted to the solvent and paper ✓

Remember
To find the number of substances count the number of spots.

Remember
Each substance can be identified by how far it travels up the chromatography paper.

Remember
To give reasons when you are asked to explain things. Include connectives like because.

5. Harry thinks that black ink is a mixture but Khalid disagrees.

a Describe how the boys could use chromatography to find out who is right.

First draw a _____ just above the bottom of a strip of filter paper.

Next put a small spot _____ on the line.

Place the paper in a beaker of solvent so that the _____ is just above the solvent.

Then leave it for about 40 minutes to see if a _____ .

b The boys set up their experiment and here is the result.

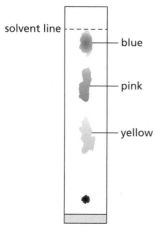

solvent line — — — —
— blue
— pink
— yellow

Who is correct, Harry or Khalid?

Give a reason for your answer

_____

_____

_____

_____

6. Chromatography can be used to test for banned substances such as performance-enhancing drugs sometimes found in the bloodstream of athletes.

The image shows the chromatograms of four banned substances (BS) next to the blood sample taken from an athlete.

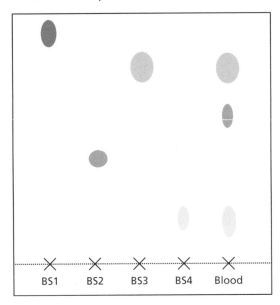

a Write down the number of banned substances found in the blood sample.

_____

b Give a reason for your answer to (a).

_____

c Use evidence from the chromatogram to explain whether the blood sample contains any banned substances.

_____

_____

_____

_____

d What further evidence could be collected in order to support the answer given in part (c)?

_____

_____

_____

_____

7.

Practical

You might like to have a go at making your own chromatogram at home. You could test the inks from colour pens or the dyes from sweets. Instead of chromatography paper you could try to use a coffee filter paper or kitchen roll.

Self-assessment

Tick the column which best describes what you know and what you are able to do.

What you should know:	I don't understand this yet	I need more practice	I understand this
Rutherford's model was that an atom has a small, central, positively charged nucleus, but most of it is empty space			
Electrons have a negative charge, protons have a positive charge and neutrons have no charge			
The electrostatic attraction between positive and negative charges holds together individual atoms			
Chromatography is used to separate mixtures of soluble substances			
The final pattern created is called a chromatogram			
During chromatography soluble dyes separate out at different locations			
Chromatography can be used to identify unknown substances in a mixture			

You should be able to:	I can't do this yet	I need more practice	I can do this by myself
Make predictions of likely outcomes based on scientific knowledge			
Describe an analogy and how it can be used as a model			
Make conclusions by interpreting results			
Explain the limitations of the conclusions			

If you have ticked 'I don't understand this yet' or 'I can't do this yet' or mostly 'I need more practice', have another look at the relevant pages in the Student's Book. Then make sure you have completed all the questions in this Workbook chapter and the review questions in the Student's Book. If you have already completed all the questions, ask your teacher for help and suggestions on how to progress.

Teacher's comments

End-of-chapter questions

1. Look at the diagram of a nitrogen atom.

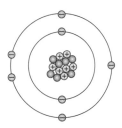

a How many electrons are in the atom? _____

b How many neutrons are in the atom? _____

c How many protons are in the atom? _____

d Write down the chemical symbol of nitrogen. _____

e Complete the table.

Particle	Charge
Proton	
	Negative
Neutron	

f Describe how individual atoms are held together.

_____

_____

2. Charlie wants to investigate a mixture of amino acids to find out which amino acids are present.

a Describe how Charlie could use chromatography in his investigation.

_____

_____

_____

_____

The diagram shows the resulting chromatogram when Charlie tested the unknown mixture against four known amino acids.

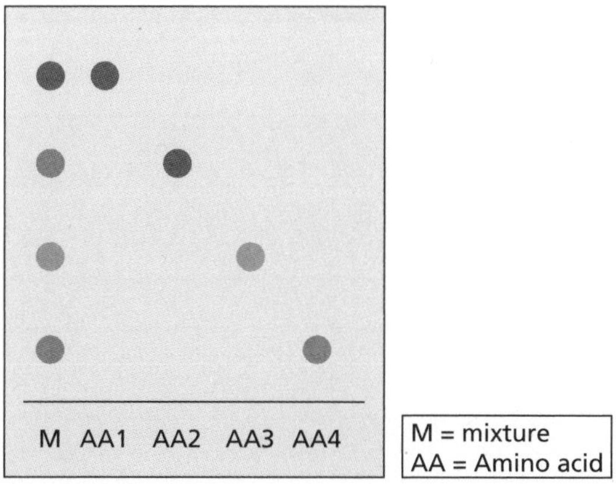

M AA1 AA2 AA3 AA4

M = mixture
AA = Amino acid

b Explain why Charlie also tested some known amino acids.

_____

_____

c How many substances were in the mixture? _____

d What conclusion can Charlie make about the amino acids present in the mixture?

_____

_____

_____

3. Bhakti and Vimla were using chromatography to investigate black felt tip pens.

Here are the results:

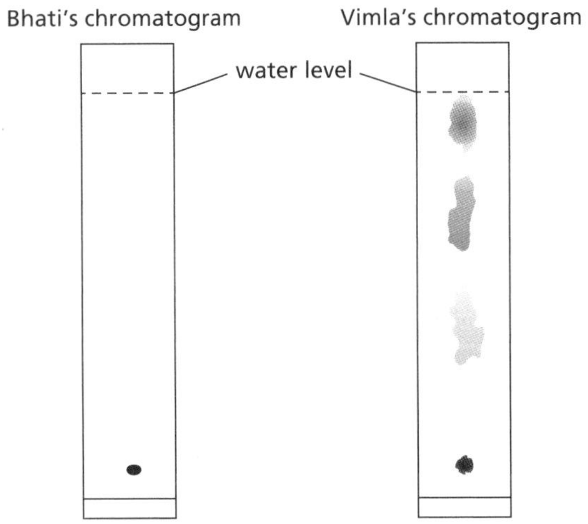

Bhati's chromatogram   Vimla's chromatogram

water level

a Compare the two chromatograms.

_____

_____

b Suggest why the chromatograms look different.

_____

_____

c   Explain how chromatography works.

_____

_____

_____

_____

4.   Scientists use models to try and explain what is inside the atom.

a   Why do you think scientist use models to try and explain what is inside the atom?

_____

_____

b   Suggest a reason why models change over time.

_____

_____

c   In 1909 Ernest Rutherford came up with a new model of the atom.

Describe the main features of Rutherford's model.

_____

_____

_____

_____

d   Describe how Rutherford's model was different to the well-established 'plum pudding' model of the atom.

_____

_____

5.1 Concentration of solutions

You will learn:

- To understand what is meant by the concentration of a solution
- To choose experimental equipment and use it correctly
- To plan investigations, including fair tests, while considering variables

1. Draw a line to match the key word to its meaning. The first one has been done for you.

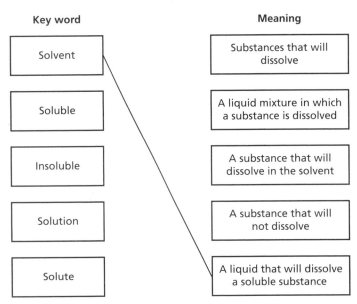

Key word	Meaning
Solvent	Substances that will dissolve
Soluble	A liquid mixture in which a substance is dissolved
Insoluble	A substance that will dissolve in the solvent
Solution	A substance that will not dissolve
Solute	A liquid that will dissolve a soluble substance

2. The particle diagrams show different concentrations of the same solutions.

Key   ● Solvent particles   ● Solute particles

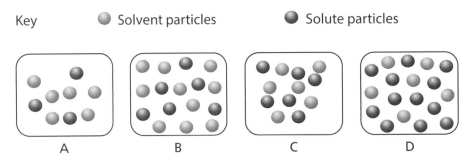

A   B   C   D

a Identify the most concentrated solution.

Tick **one** box

A ☐   C ☐

B ☐   D ☐

b Give a reason for your answer to part (a).

_____

_____

3. Adi is making a vitamin C drink. He added a vitamin C tablet to a glass of cold water and stirs. Soon the water has gone completely orange.

Worked Example

a Name the solvent _water._

b Name the solute _ vitamin C tablet._

c Complete the diagrams to show what happens to the particles during dissolving.

Key   ● Water particles   ● Vitamin C tablet particles

at the start   during   at the end

d Suggest a reason why the water looked orange at the end.

The vitamin C particles are orange. As the vitamin C dissolves, the particles spread out into the water. By the end the water looks orange because it is now a solution. The orange vitamin C particles are everywhere.

4. Gonzalo is making a drink for his friend Jose. He adds a teaspoon full of blackcurrant crystals into a beaker of water and stirs.

a Name the solvent. _____

b Name the solute. _____

c When Jose tries the drink he says that it doesn't really taste of anything.

What should Jose do to improve the taste? Give a reason for your answer.

_____

_____

Marco is making up some salt solutions using the amounts given in the table.

Solution	Volume of water in cm³	Mass of salt in g
A	20	5
B	20	10
C	10	4
D	10	1

a Name the piece of equipment Marco should use to measure:

i The water _____

ii The salt _____

b Which salt solution is the most concentrated?

Circle **one** letter.

A   B   C   D

c Which salt solution is the most dilute?

Circle **one** letter.

A   B   C   D

d Complete the particle diagrams.

Key   ● Water particles   ● Salt particles

at the start   during   at the end

e When Marco heats up the water, he finds that more salt will dissolve in it.

Explain why.

_____

_____

6. A group of students were investigating the solubility of baking powder at 20 °C.

Practical They added 1 g of baking powder to 100 g of water and stirred until it dissolved.

This was repeated until no more baking powder would dissolve.

The results are presented in the table.

Mass of baking power (g)	All dissolved ✓ or ✗
1	✓
2	✓
3	✓
4	✓
5	✓
6	✓
7	✓
8	✓
9	✓
10	✗
11	✗
12	✗

a Name the piece of equipment the students should use to measure out the baking powder.

_____

b Identify two variables that must be controlled.

_____

c At what point did the solution become saturated?

_____

d What further evidence could the group of students collect to improve the accuracy of their answer to part (c)?

_____

_____

e Use your ideas about particles to describe what happens when a solution becomes saturated.

_____

_____

5.2 Solubility

You will learn:

- To describe the effects of temperature on the solubility of salts
- To plan investigations, including fair tests, while considering variables
- To describe results in terms of any trends and patterns and identify any abnormal results
- To reach conclusions studying results and explain their limitations
- To present and interpret scientific enquiries correctly

1. Complete the sentences using words from the box.

Temperature   solution   chemical   solvent

soluble   solute   physical   pressure

Solubility is a _____ property of a substance. It is the mass of

_____ that will dissolve in a volume of _____ at a

certain _____ .

2. Magnesium sulfate may be used in medicines. At 20 °C it has a solubility of 35.1 g per 100 cm³ water.

Complete the sentence.

At a temperature of 20 °C _____ magnesium sulfate will dissolve

in _____ water.

3. The solubility of some salts in 100 cm³ of water at 20 °C are listed below.

Worked Example

Potassium sulfate 11.1g

Ammonium nitrate 192g

Barium chloride 35.8g

Zinc carbonate 4.7×10^{-5} g

   a   Which salt is the most soluble? ____ *Ammonium nitrate* ✓ _____ .

   b   Give a reason for your answer to part (b).

      More ammonium nitrate will dissolve than any of the other salts ✓

   c   Which salt could be described as insoluble? *Zinc carbonate* ✓

d Give a reason for your answer to part (b).

Only a very tiny amount of zinc carbonate will dissolve in 100 cm³ water ✓

4. The table shows the solubility of some salts in 100 cm³ of water at different temperatures.

Salt	Solubility in g in 100 cm³ water at different temperature		
	20 °C	40 °C	60 °C
Calcium sulfate	0.26	0.27	0.24
Potassium nitrite	306.0	329.0	348.0
Sodium chloride	35.9	36.4	37.0

 a Describe how the temperature affects the solubility of potassium nitrate.

As the temperature increases from 20 °C to 60 °C the solubility _____

_____ .

b Predict the solubility of sodium chloride at 80 °C. _____

c Identify the anomalous result. Draw a circle around the number in the table.

Remember
An anomalous result is one that does not fit the pattern.

d Give a reason for your answer.

_____

_____

5. The particle diagram represents the solubility of sodium chloride at 20 °C.

Key
- sodium chloride particles
- water particles

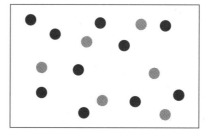

Explain how the diagram would change to represent the solubility of sodium chloride at 50 °C.

Remember
When you explain something you must say how it will change and give a reason for the change. You may include a particle diagram in your answer.

_____

_____

_____

_____

6. A fish breathes underwater through its gills. Dissolved oxygen in the water moves through the thin walls of the gills into the blood where it can travel to the fish's cells.

The graph shows how the solubility curve for oxygen in mg per litre of water.

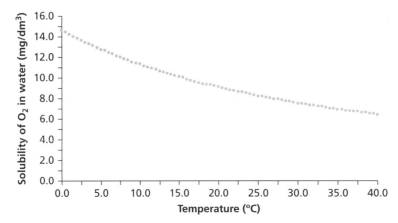

a Describe how the temperature affects the solubility of oxygen.

_____

_____

Challenge **b** Use the particle model to suggest a reason for this.

_____

_____

c Tropical fish thrive in water temperatures of around 25 °C. Suggest a reason why.

_____

_____

7. A group of students were investigating the solubility of ammonium nitrate. At each temperature they added ammonium nitrate to 100 g of water until no more would dissolve.

Practical The results are presented in the table.

Temperature (_____)	Mass of ammonium nitrate (g/100 g water)
0	30
20	28
40	46
60	59
80	66
100	78

a Write the unit of temperature onto the table.

b Write down the independent variable.

_____

c Write down the dependent variable.

_____

d Identify a variable that must be controlled.

_____

e Label the x and y axes.

Plot the data points on to the graph.

Draw a line of best fit.

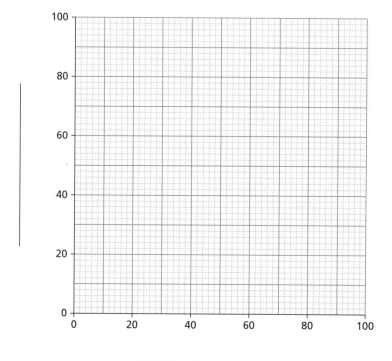

_____

> **Remember**
>
> When plotting a point on a graph 'go along the corridor and up the stairs' i.e. count along the x axis first and then the y axis.

f Identify the anomalous result. _____

g How could the group improve the quality of their results?

_____

h What can conclusions can you draw from the graph?

_____

ⓘ Use your ideas about particles to describe what happens when ammonium nitrate dissolves in water.

_____

_____

Self-assessment

Tick the column which best describes what you know and what you are able to do.

You should know:	I don't understand this yet	I need more practice	I understand this
The concentration of a solution relates to how many particles of the solute are present in a volume of the solvent			
A dilute solution has a small number of solute particles dissolved in a large volume of solvent			
A concentrated solution has a large number of solute particles dissolved in a small volume of solvent			
The solubility of salts varies, depending on their type			
The solubility of most solutes increases with temperature			

You should be able to:	I can't do this yet	I need more practice	I can do this by myself
Use an existing analogy for another purpose			
Plan an investigation			
Describe the trends and patterns in results, including identifying any anomalous results			
Present measurements appropriately, using graphs			

If you have ticked 'I don't understand this yet' or 'I can't do this yet' or mostly 'I need more practice', have another look at the relevant pages in the Student's Book. Then make sure you have completed all the questions in this Workbook chapter and the review questions in the Student's Book. If you have already completed all the questions, ask your teacher for help and suggestions on how to progress.

Teacher's comments

End-of-chapter questions

1. The particle diagrams show different samples of orange squash.

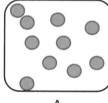 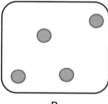 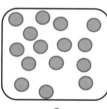 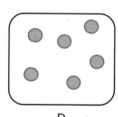

| A | B | C | D |

a Put the samples in order of increasing concentration.

_____

b Eliza tasted the most concentrated sample and found she didn't like it because it was too weak.

Describe how she could improve the drink. Give a reason for your answer.

_____

_____

2. The table shows the solubility of potassium nitrate at different temperatures.

Temperature (°C)	Solubility (g/100g water)
20	32
40	64
60	110
80	169
100	246

a Label the x and y axis on the graph below.

b Plot data points on to the graph.

c Draw the solubility curve for potassium nitrate by joining up the points with a smooth curve.

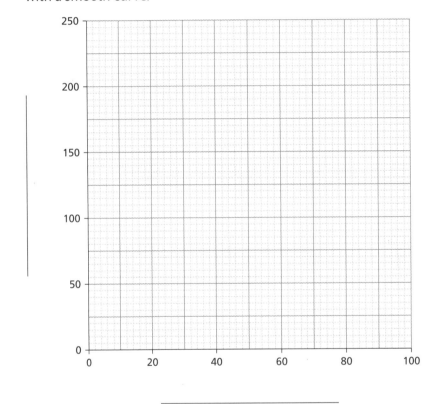

d Describe the pattern shown by the graph.

_____

_____

e What is the solubility of potassium nitrate at 50 °C? _____

f How much potassium nitrate will dissolve in 50 cm³ of water at 40 °C?

Show your working.

_____

_____

3. Tommy and Erasmus want to compare the solubility of two different salts A and B. They collect together the following equipment:

Balance

Beaker

Spatula

Measuring cylinder

Stirring rod

a Describe how the boys plan to use the equipment to measure the solubility.

_____

_____

_____

_____

b Describe how the boys will present their results.

_____

_____

Challenge **c** Use your knowledge of the particle model to explain why different salts have different solubilities.

_____

_____

_____

_____

6.1 Using word equations

You will learn:
- To describe reactions using word equations
- To understand what unreactive substances are
- To represent scientific ideas using symbols and formulae
- To reach conclusions studying results and explain their limitations

1. A word equation is used to describe a chemical reaction.

Identify the correct way of writing a word equation.

Tick **one** box.

☐ reactants = products   ☐ reactants → products

☐ products ← reactants   ☐ products → reactants

2. Elements and compounds are made from atoms.

Identify the element.

Tick **one** box.

☐ H_2SO_4 ☐ S_8 ☐ H_2O ☐ SO_2

3. Some substances are inert.

Identify the inert substance.

Tick **one** box.

☐ oxygen   ☐ neon   ☐ hydrochloric acid   ☐ iron

4. The total mass of a beaker of milk and a beaker of orange juice is 125 g.

Practical

When the milk is poured into the orange juice, a cream-coloured solid is formed but the mass stays the same.

Show Me

Explain the observations.

When the orange and milk was mixed a _____

took place. New compounds were _____

formed as the mass stayed the same because the

_____ provided all the atoms

needed to make the products.

Remember

Questions of this type find out if you can apply your knowledge of chemical reactions to everyday materials.

- milk
- orange juice
- balance

5. The particle diagram shows the atoms present in some reactants.

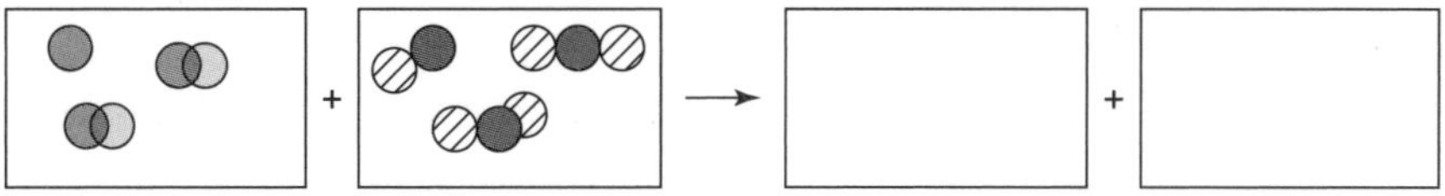

During the reaction, the atoms rearranged to form new compounds.

Show the atoms present in the new compounds by drawing in the last two boxes.

6. Sasha is getting ready for her sister's birthday party. She wants to decorate the house with some 'floating' balloons.

Sasha wants to fill the balloons with hydrogen gas but her brother says that hydrogen is unsafe, she must use helium gas.

Explain why Sasha's brother is correct.

Remember
When writing a word equation, you will find the answer in the question.

_____

_____

7. When you add sodium hydroxide to iron(II) nitrate, you can see a green solid (iron(II) hydroxide) forming in colourless sodium nitrate solution.

Show Me

Write a word equation for the reaction.

Remember
When writing a word equation, write the reactants; then draw an arrow and write the products. Read the question carefully before attempting the equation.

sodium hydroxide + _____ →

iron(II) hydroxide + _____

8. Farmers often put lime on fields to neutralise acidic soil.

This is the label usually found on a bag of lime.

State two things that the label tell us about lime.

CaO

_____

_____

9.

Challenge

Magnesium metal burns with a bright white light. At the end of the reaction, white magnesium oxide powder is produced.

Suggest why magnesium is often found in fireworks. Your answer should include a word equation.

_____

_____

_____

10.

Practical

Eddie was carrying out a series of experiments. He has noted down his observations in a table.

Test	Method	Observations	Conclusion
1	Add 1 drop of sodium hydroxide to iron(III) sulfate	A rust-coloured solid appeared	
2	Hold a lighted splint to a test tube of helium gas	No change observed	
3	Heat copper metal in a flame	The metal went black	
4	Add copper metal to hydrochloric acid	No changes observed	
5	Add magnesium metal to hydrochloric acid	Bubbles seen coming off the metal	

a Complete the table.

Write a conclusion for each experiment based on the observations.

b Write a word equation for tests 1, 3 and 5. You may need to look up the names of some of the products if you cannot work them out.

Test 1:

_____ + _____ →

_____ + _____

Test 3:

_____ + _____ →

_____

Test 5:

_____ + _____ →

_____ + _____

6.2 Pure substances and mixtures

You will learn:
- To describe correctly the purity of a mixture
- To know that reactions can produce single pure products or impure mixtures
- To identify and control risks for practical work

1. Look at the particle diagrams.

A   B   C   D

a Which two diagrams show pure substances? _____

b Which diagram shows a mixture of compounds? _____

2. Draw **one** line to match the key word with its meaning.

The first one has been done for you.

Key word	Meaning
Mixture	Two or more elements or compounds mixed together
Purity	Substances containing only one element or compound
Pure	A mixture of metals with other elements
Alloy	How much of a chemical is in a mixture

3. Complete the sentences using words from the box.

risk	assessments	control	hazards

_____ are things that can cause harm. A _____ is the

chance of a hazard causing harm to you or the people around you. Scientists write

risk _____ to help them identify and _____ risks.

4. During the cold winter season, salt is sprinkled onto the roads to stop ice forming on the surface.

Worked Example

Explain why.

The salt mixes with the water making it impure ✓ The impure salty water has a lower melting point which stops the water freezing to form ice ✓

5. Ammonia is used to make fertilisers. It is produced when nitrogen and hydrogen react together. The particle diagrams show the reaction.

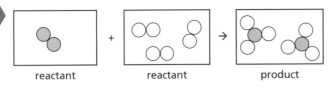

| reactant | reactant | product |

The purity of the ammonia produced is 15%.

Draw particle diagrams to show what substances could be making the product impure.

In some chemical reactions not all the reactants are used up. Look at the particle diagram of the reaction to find the answer.

_____

6. Kaleb and Titus are going to make some copper sulfate crystals by mixing sulfuric acid and copper oxide but first they must complete a risk assessment.

a Explain why it is important for the boys to carry out a risk assessment?

_____

_____

b Complete the risk assessment.

Substance	Hazard	Risk	Controlling the risk
Dilute sulfuric acid	⚠ Irritant	It may irritate the eye and skin	
Copper oxide	⚠ Irritant		
	Environment (toxic to aquatic wildlife)		Do not put down the sink. Leave for technician to dispose.

The word equation for the reaction is:

Sulfuric acid + copper oxide → copper sulfate + water

c   Describe how Kaleb and Titus will obtain pure copper sulfate crystals once the reaction is complete.

_____

_____

7.   The purity of gold is measured in karats. Pure gold is 24 karat.

9 karat gold is often used to make jewellery such as earrings and necklaces.

a   Suggest a reason why.

_____

b   The particle diagram represents 24 karat gold.

Draw a particle diagram in the box below to represent 12 karat gold.

Challenge   c   Gabriella has a 4 karat gold bracelet. It has a mass of 12g.

Calculate the mass of gold in the bracelet.

Show your working.

_____

_____

6.3 Measuring temperature changes

You will learn:

- To use temperature change to identify chemical reactions
- To evaluate experiments and investigations, and explain any improvements suggested

1. Which device uses a chemical reaction to produce a decrease in temperature?

Tick **one** box.

☐ Self-heating can   ☐ Instant cold pack

☐ Hand warmers   ☐ Car engine

2. Thermometers are used to measure temperature.

What temperature is this one showing? _____

3. Complete the table.

Worked Example

Reaction	Start temperature (°C)	Final temperature (°C)	Temperature difference (°C)
A + B	19	27	27 − 19 = +8
C + D	18	11	11 − 18 = −7

72

State which reaction is exothermic.
Explain how you can tell.

A + B is exothermic because there is an
increase in temperature.

4. Safia and Mia mixed some citric acid with sodium
hydrogencarbonate solution.

The girls recorded the temperature of the mixture at the start
of the reaction. They also recorded the lowest temperature of
the reaction mixture.

- Starting temperature 20 °C

- Lowest temperature 7 °C

Describe and explain what happened during the reaction.

The temperature decreased by _____ , so _____ was

transferred from the surroundings during the reaction as heat.

5. Youssef was investigating the temperature changes that take place when magnesium
ribbon is added to dilute hydrochloric acid. Here are his results.

Length of magnesium (cm)	Temperature at start (°C)	Temperature at end (°C)
1.0	20.1	20.9
2.0	20.2	21.6
3.0	19.9	22.3
4.0	20.0	23.1

State the conclusions Youssef can make from this data.

_____

_____

_____

6.

Practical

Yun and Safia are investigating the energy changes that take place when calcium and dilute hydrochloric acid are mixed together. During the investigation the students record the temperature at the start and the end of each reaction.

Figure **X** is a diagram of Yun's apparatus and Figure **Y** is a diagram of Safia's apparatus.

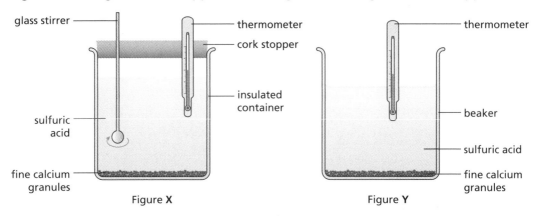

Figure **X**   Figure **Y**

a   Compare the two sets of apparatus. List **two** similarities and **two** differences.

_____

_____

_____

_____

b   Explain why Yun used a cork stopper.

_____

c   Explain which set of apparatus (Yun or Safia's) will give the most accurate results. Give a reason for your answer.

_____

_____

d   Name **two** other pieces of apparatus they will need to use to make sure their results are accurate. Explain your answers.

_____

_____

6.4 Exothermic and endothermic processes

You will learn:

- To use temperature change to identify endothermic and exothermic processes

1. Which of the following is an example of an endothermic process?

   Tick **one** box.

   ☐ Freezing   ☐ Condensation

   ☐ Photosynthesis   ☐ Combustion

2. Complete the sentences that follow by choosing words from this list.

increase	decrease	endothermic	light
exothermic	combustion	heat	energy

   _____ processes transfer energy to the surroundings – often as

   heat – and cause a temperature _____ .

   _____ processes transfer energy from the surroundings, often

   as _____ .

3. Draw lines to match each change of state with its energy change.

   Change of state   **Energy change**

   | Melting | | Exothermic |

   | Freezing | | Endothermic |

   | Condensing |

   | Evaporating |

   Remember
   When answering an 'explain' question you need to give a reason for your answer.

4. Explain how sweating cools the skin.

Show Me

Energy transfers from the skin to sweat as _____ .

The sweat _____ leaving the skin cooler.

5. Priya and Lily are investigating the energy changes that take place during dissolving.

Challenge They follow this method:

- Pour 10 cm³ of water into a boiling tube.

- Record the temperature of the water.

- Add 0.5 g of ammonium chloride and stir.

- Record the temperature when the ammonium chloride has dissolved.

Here are their results:

Start temperature in °C   20.0

Final temperature in °C   18.5

Priya thinks that if they repeat the experiment using 9 cm³ water the temperature change will stay the same, but Lily thinks it will decrease further.

a Explain why Lily is correct.

_____

_____

b Evaluate the method and identify any improvements you would make.

_____

_____

6. The word equation for photosynthesis is:

Challenge water + carbon dioxide → glucose + oxygen

The word equation for respiration is:

glucose + oxygen → carbon dioxide + water

Use your scientific knowledge to explain which process is exothermic and which is endothermic.

_____

_____

_____

7.

Practical

Oliver and Mike put some ice into a container and heat it.

They record the temperature every minute. This graph shows their results.

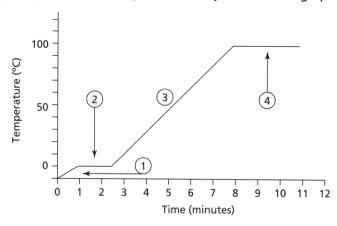

Use the graph to answer these questions.

a What was the temperature of the ice at the start? _____

b At what temperature did the ice melt? _____

c Explain what is happening to the particles during step 2.

_____

_____

d Name the type of energy change taking place during step 4. Give a reason for your answer.

_____

_____

6.5 The reactivity series

You will learn:

- To describe the reactivity of metals with oxygen, water and dilute acids
- To plan investigations, including fair tests, while considering variables
- To identify and control risks for practical work
- To describe results in terms of any trends and patterns and identify any abnormal results
- To present and interpret scientific enquiries correctly
- To reach conclusions studying results and explain their limitations

1. Look at the diagram. It shows some metals reacting in water.

Put the metals in order of reactivity, starting with the most reactive.

Tick **one** box.

[ ] A B C D [ ] A D C B [ ] B A D C [ ] D B C A

2. Which of the following metals shows the fastest reaction with cold water?

Tick **one** box.

[ ] Copper [ ] Sodium [ ] Zinc [ ] Magnesium

3. Choose words from the list to complete the sentences that follow.

| reactive   gold   silver   unreactive   oil   water   acid   middle   bottom   top |

The Californian gold rush began in 1848 when _____ was found

by James Marshall at Sutter's Mill. Gold is very _____ or inert.

It does not react with _____ or oxygen, which is why it is

sometimes found in river beds. Gold is found towards the _____

of the reactivity series of metals.

4.

Practical

A teacher carefully cut some samples of metals and left them in the air. The class timed how long it took for the cut edge of the metals to go dull and recorded the results on scrap paper. The times were:

potassium: 13 s   calcium: 2 min 30 s   sodium 57 s.

a   Put the data into this blank table. Start by giving each column a heading.

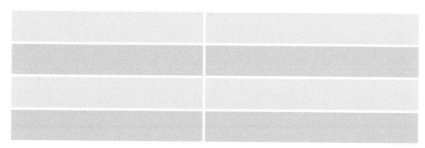

b   Underline the metal showing the fastest reaction.

c   Draw a circle round the slowest reaction.

5.

Worked Example

When metals react with dilute acids, a salt is formed and hydrogen gas is given off.

Draw a particle diagram to show the reaction between zinc and sulfuric acid.

Start by writing the word equation.

Use the information in the question to help.

Zinc + sulfuric acid → zinc sulfate + hydrogen ✓

Draw the atoms on both sides of the equation. You must draw enough particles to represent all the 'reacting' parts of the equation. Here not all the individual particles in the sulfate group have been shown.

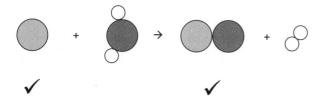

✓   ✓

6.

When calcium reacts with water, the reaction is:

- more vigorous (faster) than the reaction between zinc and water

- less vigorous (slower) than the reaction between magnesium and water.

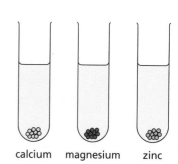

calcium   magnesium   zinc

a Predict how the metals will react in hydrochloric acid by drawing bubbles in the test tubes.

_____

_____

Show Me

b Draw a particle diagram to show the reaction between zinc and water.

The word equation is:

| Zinc | + | water | → | zinc hydroxide | + | hydrogen |

| | | | | | | |

7. Aluminium is above iron in the reactivity series of metals.

Challenge Observations show that objects made from aluminium do not corrode as quickly as those made from iron. Use your research to explain why aluminium corrodes less quickly than expected to explain the observations.

_____

_____

_____

8. Look at the diagram of the reactivity series – it shows the metals in order of reactivity.

Read these statements:

- Copper beads over 10 000 years old have been found in Northern Iraq.

- The ancient Hittites of Asia Minor extracted iron in the 1500s Before the Common Era (BCE).

- Sodium metal was first extracted in 1807.

Most reactive

K	potassium
Na	sodium
Ca	calcium
Mg	magnesium
Al	aluminium
C	carbon
Zn	zinc
Fe	iron
Sn	tin
Pb	lead
H	hydrogen
Cu	copper
Ag	silver
Au	gold
Pt	platinum

Least reactive

a Describe the pattern you can see between a metal's place in the reactivity series and its discovery.

_____

b Suggest an explanation for the pattern.

_____

_____

_____

_____

_____

9. **Investigating the reactivity of metals**

Practical Priya and Carlos are investigating the reactivity of metals with different liquids.

Experiment 1: They add small pieces of the metals to water.

Experiment 2: They add small pieces of the metals to dilute hydrochloric acid.

a What two variables should Priya and Carlos control?

_____

_____

b Describe two safety precautions they should take.

_____

_____

c Here are the results of the experiments.

Metal	Experiment 1 Reaction with water	Experiment 2 Reaction with hydrochloric acid
A	Lots of bubbles	Lots of bubbles forming very quickly
B	No reaction	Some bubbles on surface of metal
C	No reaction	No reaction
D	A few bubbles on surface of metal	Lots of bubbles

Write the metals in order of reactivity, starting with the most reactive.

_____

d State which metal could be copper. Give a reason for your answer.

_____

_____

e State which metal could be calcium. Give a reason for your answer.

_____

_____

f Write a word equation for the reaction of calcium with water.

_____

g Complete the particle diagram for the reaction of calcium with hydrochloric acid.

h The students decide to repeat the experiment using dilute sulfuric acid.

Predict what will happen when some sulfuric acid is added to metal **A** and to metal **C**.

_____

_____

Self-assessment

Tick the column which best describes what you know and what you are able to do.

You should know:	I don't understand this yet	I need more practice	I understand this
During a chemical reaction, reactants react to form new products			
A word equation separates the substances that react (on the left) and the products that are formed (on the right) with an arrow			
A pure substance only contains one type of particle			
The purity of a mixture tells us how much chemical is in a mixture			
Reactions do not always result in a single pure product. Often they produce a mixture of products			
During chemical reactions, energy can be transferred to or from the surroundings.			
The temperature of thermal (heat) energy to or from the surroundings causes the temperature to change			
Energy is transferred to the surroundings by exothermic processes and chemical reactions, causing the temperature of the surroundings to increase			
Energy is transferred from the surroundings by endothermic processes and chemical reactions, causing the temperature of the surroundings to decrease			
Metals react in similar ways, but some are more reactive than others			
Information about the reactivity of metals with oxygen, water and dilute acids is used to put them into the reactivity series			

You should be able to:	I can't do this yet	I need more practice	I can do this by myself
Use symbols to represent chemical formula			
Make risk assessments for practical work to identify and control risks			
Plan an investigation using previous knowledge and understanding			
Evaluate an investigation, suggesting improvements and explaining any planned changes			
Make predictions of likely outcomes based on scientific knowledge and understanding			
Explain why accuracy and precision are important			
Make conclusions by interpreting results			
Plan a range of investigations, while considering variables appropriately and how many repetitions are needed to be reliable			
Decide on what equipment is required to carry out an investigation			

If you have ticked 'I don't understand this yet' or 'I can't do this yet' or mostly 'I need more practice', have another look at the relevant pages in the Student's Book. Then make sure you have completed all the questions in this Workbook chapter and the review questions in the Student's Book. If you have already completed all the questions, ask your teacher for help and suggestions on how to progress.

Teacher's comments

End-of-chapter questions

1. Mrs Brown's class were investigating how the mass of magnesium changes when it burns.

   a Write a word equation for the reaction.

   _____ + _____ → _____

   At the end of the experiment they collected the results together in a table:

Group	Mass of magnesium at start	Mass of product at end
1	0.51	0.65
2	0.52	0.72
3	0.50	0.35
4	0.51	0.75

   b State what is missing from the table.

   _____

   c Explain which group's result was unexpected. Give a reason for your answer.

   _____

   d Describe what the class could do to try and avoid unexpected results.

   _____

2. This question is about chemical reactions with acids.

   a When calcium is added to hydrochloric acid, a gas is formed.

   Name the gas formed.

   Tick **one** box.

   ☐ Oxygen   ☐ Hydrogen

   ☐ Carbon dioxide ☐ Sodium hydroxide

   b Describe an observation you could make which would tell you that a gas has been formed.

   _____

   _____

c Describe what happens to the atoms during the reaction in (b).

You may wish to include a particle diagram.

_____

_____

3. Chen is investigating endothermic and exothermic reactions. He mixes different chemicals together and records the temperature changes.

a Complete the table of results.

Chemicals	Start temperature (°C)	Final temperature (°C)	Temperature change (°C)	Endothermic or exothermic
Sodium hydroxide and sulfuric acid	18	21	+3	
Ammonium chloride and water	18	13	−5	
Magnesium and hydrochloric acid	18	25		

b Predict what will happen to the temperature change if Chen increases the mass of magnesium used but keeps all other conditions the same.

_____

c Chen is wanting to make a cool pack to keep his lunch fresh.

i Which chemicals could he use?

_____

ii Describe **two** things he could do to make the pack more effective.

_____

_____

4. Two students were investigating the reactivity of metals with sulfuric acid.

a Identify a hazard and suggest a control measure that could be taken to minimise the risk.

_____

_____

The results are recorded in the table.

Metal	Observation when sulfuric acid was added
cobalt	A few bubbles
magnesium	Vigorous bubbles
copper	No change
zinc	Lots of bubbles

b Determine the order of reactivity of the metals.

Put them in order, starting with the most reactive.

_____

c Suggest a reason for your answer to part b.

_____

d Cobalt reacts with sulfuric acid to produce a salt and hydrogen gas. Write the word equation for this reaction.

_____ + _____ → _____ + _____

e The experiments were repeated using hydrochloric acid. Predict the order of reactivity.

_____

5. Gabriella wants to find out which hydrocarbon fuel releases the most energy when it is burned.
She heats equal volumes of water using a known mass of each fuel.

Practical She sets up the apparatus shown in the diagram on the right.

a Complete the diagram by adding the missing two labels.

b During the investigation, Gabriella records some data in a table.

Write in the missing headings in the table.

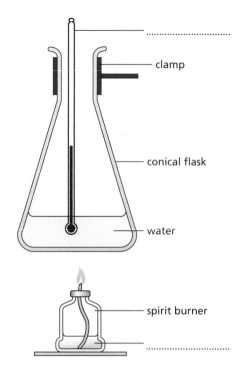

....

clamp

conical flask

water

spirit burner

....

Name of fuel	Mass of fuel at start (g)			Temperature of water at end (°C)

c Look at the diagram. Does all the heat produced go into raising the temperature of the water? Give a reason for your answer.

_____

_____

d After completing her experiments Gabriella decides to extend her investigation to include wood, which is a solid. Describe how she should modify her apparatus.

_____

_____

Physics

7.1 Measuring distance and time

You will learn:

- To plan investigations, including fair tests, while considering variables
- To choose experimental equipment and use it correctly
- To take accurate measurements and explain why this matters
- To evaluate the reliability of measurements and observations

1.

Practical

Complete the sentences using words from the list.

speed	**force**	**three**	**distance**	**time**	**mass**	**mean**
		weight		**anomalous**		

We can use a ruler or measuring tape to measure _____ .

We can use a stop clock or light gates to measure _____ .

We need to measure both these things to calculate an object's _____ .

To make sure our results are reliable we need to: Do the experiment at least

_____ times. Remove any _____ results and calculate the

_____ .

2. Look at the diagram of four lengths we can measure. Which measuring device is
the most accurate for measuring each length? Choose the **best** device from the list.

- 15 cm ruler marked in millimetres
- metre rule marked in millimetres

- 10 m measuring tape marked in centimetres
- 100 m measuring tape marked in centimetres

A

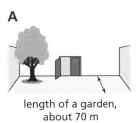

length of a garden,
about 70 m

B

width of a box,
about 10 cm

C

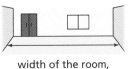

width of the room,
about 6 m

D

length of a person's stride,
about 80 cm

A:_____

B:_____

C:_____

D:_____

3.

Practical

Hassan wants to measure accurately how far he can jump. He chooses a 30 cm ruler marked in millimetres.

a Explain to Hassan why this is not the best choice of measuring device.

Worked Example

Hassan can probably jump about 2 m. He would need to place the ruler end-over-end about 7 times to measure the distance. ✓ This makes it likely he would make errors. ✓

b Suggest to Hassan a better choice of measuring device. Explain your answer.

A measuring _____ marked in _____ . This will be more accurate because it only needs to be placed

_____ .

> **Remember**
>
> You also need to match the accuracy of the measuring device to the quantity being measured. Here, measuring to the nearest mm is probably not possible, so a device that measures to the nearest 1 cm or 0.5 cm is enough.

4.

Practical

Angelique uses a watch to record how long it takes to run 400 m. The first part of the diagram (**A**) shows the face of the watch.

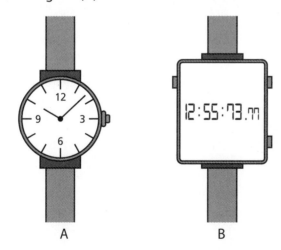

A

B

a   Would this watch give Angelique an accurate measurement? Explain your answer.

_____

_____

_____

b   The second part of the diagram (**B**) shows the face of a stop watch that Aiko uses to make the same measurement. How accurate is this watch?

_____

_____

5.

Practical

The diagram shows an experiment that Oliver has set up to measure the **time** it takes for a toy car to roll down a slope at different angles.

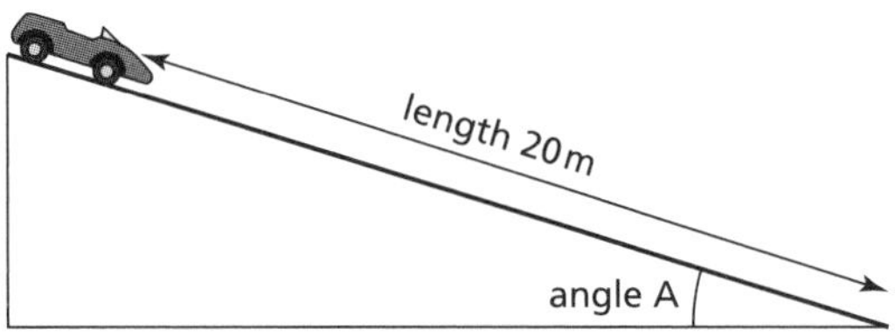

Describe the best type of device that Oliver could use for measuring the time. Include a short description of how the device works.

length 20m

angle A

_____

_____

_____

6.   Write a definition for the term 'accurate measurement'.

_____

_____

7.

Look at Oliver's experiment in question 5. The table shows the results of Oliver's experiment.

Angle of slope in degrees	Time taken in seconds	Speed
20	2.0	
30	1.7	
40	1.5	
50	0.8	

a Suggest **one** way in which Oliver could improve the accuracy of his experiment.

_____

b Time is the variable that Oliver measures. Which other variable does Oliver

need to measure to calculate the car's speed? _____

c There are other variables that might change and affect the results.
In science, we need to know about and **control** these variables to stop
them affecting the results.

Suggest how Oliver could control **two** of these variables in his experiment.

_____

_____

d How would Oliver make sure his results were reliable?

_____

_____

_____

7.2 Speed and average speed

You will learn:

- To calculate speed
- To reach conclusions studying results and explain their limitations

• •

1. Complete the sentences using words from the list.

speed	time	distance	metres	kilograms	hour	second	day

Speed describes how much _____ an object travels in a

given _____ .

The unit of speed is _____ per _____ .

2. The equation that connects average speed, distance travelled and time taken can be written in this form:

$$A = \frac{B}{C}$$

a Write down the names of the variables shown as A, B and C. Use the terms from the following list.

average speed	total distance travelled	total time taken

A = _____

B = _____

C = _____

b Explain why it is important that the total distance travelled and total time taken are used.

The totals are needed because

_____

_____

94

3. Jamila runs 20 m in 5 s. What is her average speed? Choose the **best** answer from the list.

☐ 20 m/s   ☐ 5 m/s

☐ 4 m/s   ☐ 2 m/s

4. Chen cycles at 20 m/s for 5 s. What is the distance he travels? Choose the **best** answer from the list.

☐ 20 m   ☐ 4 m

☐ 50 m   ☐ 100 m

5. 'A 100 m sprinter can run at about 10 m/s.'

Show Me

a Explain what this statement means.

A sprinter runs a race that is _____ long.

They can travel a distance of about 10 _____

in each _____ .

b Describe how a sprinter's speed changes during a race.

_____

_____

c If the sprinter's **average** speed is about 10 m/s, what can you say about the **maximum** speed the sprinter reaches?

_____

_____

6.

Practical

Pierre sets up an experiment to measure the time it takes a ball to fall to the floor from different heights. Look at Pierre's diagram of the equipment.

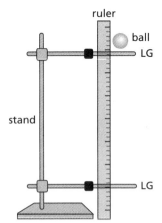

Pierre changes the height each time but always uses the same ball.

a Why does Pierre use the same ball each time?

_____

b Suggest what the measuring devices labelled 'LG' are.

_____

These are the results Pierre finds.

Height (m)	Time taken (s)	Average speed (m/s)
10	1.4	7.1
20	2.0	_____
5	1.0	_____

The first row shows the average speed for the first height. This is calculated by using the speed equation:

average speed = total distance moved / total time taken

$$= \frac{10 \text{ m}}{1.4 \text{ s}}$$

$$= 7.1 \text{ m/s}$$

c Calculate the values of speed for the second and third rows. Complete the table with these values.

d Describe the pattern in these results.

Increasing the height makes the ball's average speed_____

7. Migrating birds travel long distances. The world record for a non-stop flight by a bird is shown on the map. The bird is known as a bar-tailed godwit. It flew at an average speed of 50 km/h.

a One day contains 24 hours. How many kilometres will the godwit travel in 1 day?

_____

_____

b The non-stop flight lasted nearly 10 days. One of the distances in the list is the actual distance measured. Choose the actual distance the godwit travelled.

☐ 1170 km   ☐ 3024 km

☐ 8670 km   ☐ 11 680 km

8. The average speed of a Formula 1 racing car around the circuit at Yas Marina in Abu Dhabi is 200 km/h.

Challenge

a How many kilometres does the F1 car travel in 1 minute at this speed? Show your working.

_____

_____

Light gates are used to measure the speed of the F1 car along the main straight section of track. The record speed measured is 360 km/h.

b Explain why this measured speed is different from the average speed.

_____

_____

_____

_____

c Calculate this measured speed in metres per second, m/s. Show your working.

_____

_____

_____

_____

Remember what you learned in Stage 7 Chapter 6 about energy transfers.

d F1 cars need special brakes to slow them down. The brakes must work at very high temperatures (over 1000 °C). Use your knowledge of energy transfers and the conservation of energy to explain why the brakes must work at high temperatures.

_____

_____

_____

7.3 Distance/time graphs

You will learn:

- To interpret and draw simple distance/time graphs
- To describe trends and patterns in results
- To present and interpret scientific enquiries correctly

1. The diagram shows a distance/time graph. Complete the labels using the words from the list.

| time   speed   distance |

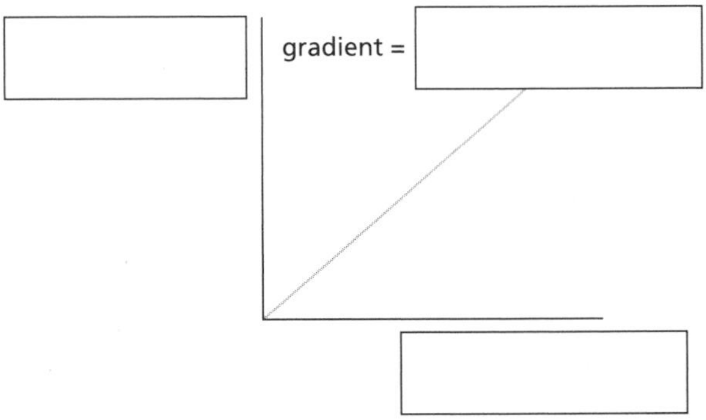

gradient =

2. The diagrams show three different distance/time graphs.

 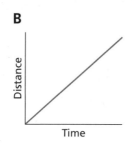 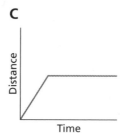

A B C

Choose the **best** graph to fit each of these descriptions.

a Car moving at a constant speed: _____

b Person running at a constant speed, then stopping and standing still:

_____

c Person standing still: _____

3. Look at the distance/time graph for a toy car.

a How far has the car moved after 4 seconds?

_____

b How long does it take the car to move a total of 6 m?

_____

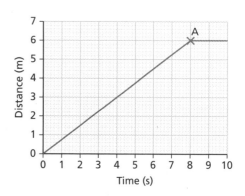

c Describe what happens at the point labelled 'A'.

The speed _____

The car _____

4. Carlos sets up an experiment to investigate a trolley rolling down a long slope. His measurements of time and distance are shown in the table.

Practical

Time (s)	Distance (m)
0	0
0.8	1.2
1.6	2.4
2.4	3.6
3.2	4.8

Look at the graph. The first three points have been placed on the graph.

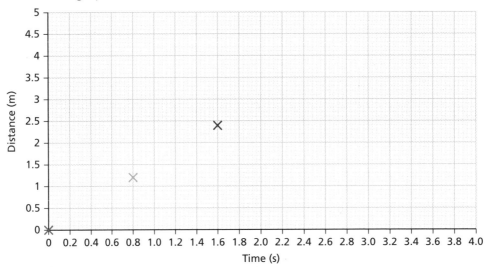

Time (s)

a Complete the graph by placing the two remaining points.

b Join all the points by drawing a line.

c Explain what the shape of the line tells us about the movement of the trolley.

_____

_____

5.

The graph shows the movement of a bus that travels between two different parts of town.

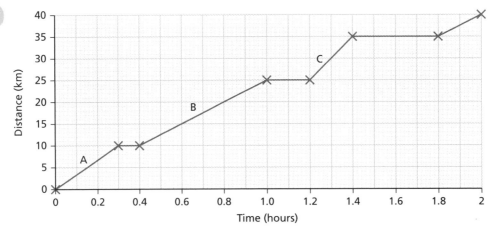

Time (hours)

a Explain what the horizontal parts of the graph tell us about the movement of the bus.

_____

_____

b Explain what the sloped parts of the graph tell us about the movement of the bus.

_____

_____

c In which section of the graph, A, B or C, is the bus moving fastest?

_____

d Calculate the **average** speed of the bus for the whole journey.

_____

_____

_____

Self-assessment

Tick the column which best describes what you know and what you are able to do.

What you should know:	I don't understand this yet	I need more practice	I understand this
Rulers and measuring tapes can be used to measure distance			
Clocks, watches, stopwatches and light gates connected to electronic timers, data loggers or computers can be used to measure time			
Different pieces of apparatus give measurements with different levels of accuracy			

	I can't do this yet	I need more practice	I can do this by myself
Average speed = distance travelled/time taken			
Speed is measured in metres per second (m/s)			
The gradient of a distance/time graph tells you the speed of a moving object			
The steeper the gradient of a distance/time graph, the faster an object is moving			

You should be able to:	I can't do this yet	I need more practice	I can do this by myself
Choose suitable apparatus to measure distance and time			
Identify important variables, choose which variables to change, control and measure			
Plan to be able to test if results are reliable			
Use a range of equipment accurately			
Present results as appropriate in tables and graphs			
Identify anomalous results from tables and graphs			
Interpret data from secondary sources			
Make simple calculations			
Identify trends and patterns in results			
Identify anomalous results and suggest improvements to investigations			
Discuss explanations for results using scientific knowledge and understanding			

If you have ticked 'I don't understand this yet' or 'I can't do this yet' or mostly 'I need more practice', have another look at the relevant pages in the Student's Book. Then make sure you have completed all the questions in this Workbook chapter and the review questions in the Student's Book. If you have already completed all the questions, ask your teacher for help and suggestions on how to progress.

Teacher's comments

End-of-chapter questions

1. Lily investigates the speed of a ball rolling along a sloped track that is 1 m long. She measures the distance using a measuring tape marked in centimetres. She uses a stopwatch to time how long the ball takes to travel along the track.

The results are shown in the table.

Repeat number	Time taken (seconds)
1	0.83
2	0.78
3	0.91

a Which measuring device could Lily use to make her measurement of distance more accurate?

_____

b Explain why using a stopwatch to measure the time is not very accurate, and suggest a better measuring device.

_____

_____

_____

c Describe how the measurements Lily has made could be used to calculate the speed of the ball.

_____

_____

2. A bus is travelling at a speed of 15 m/s. How long will the bus take to travel:

a 30 m?

_____

b 600 m?

_____

c 15 km?

_____

3. Look at the two graphs showing the movement of two different objects.

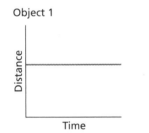
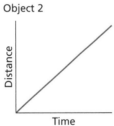

a Describe the **two** differences between the graphs.

_____

_____

b Explain what the graphs tell us about the speed of each object.

_____

_____

4. Look at the four graphs A to D.

A

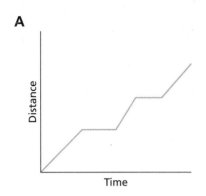

B

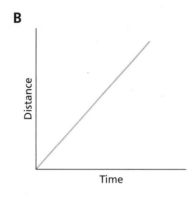

C

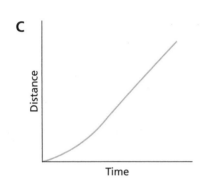

D

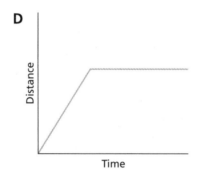

Match each graph to the description of the moving object. Choose the **best** description for each graph and write the letter of the graph next to the description.

(i) A car travelling at a steady speed. ☐

(ii) A person who walks a short distance away and then stands still. ☐

(iii) A bus that travels and makes stops along its route. ☐

(iv) An aircraft taking off, which starts stationary then speeds up until it rises into the air. After it leaves the ground, it stays at a constant speed. ☐

5. Blessy sets up an experiment to measure how long it takes for a metal ball to roll down a track. The table shows the results.

Repeat number	Time taken (seconds)
1	2.9
2	3.2
3	4.0
4	2.9

a Explain why Blessy has run the experiment four times without changing the distance travelled.

_____

_____

b Which result is anomalous (it does not fit a pattern)?

_____

c Suggest **one** reason why the anomalous result could be wrong.

_____

_____

d Choose the other **three** results. Calculate the average time taken.

_____

_____

e Blessy measures the track and finds that it is 3.6 m long. Use your answer from (d) to calculate the average speed.

_____

_____

8.1 Balanced and unbalanced forces

You will learn:
- To describe how forces affect motion
- To make predictions based on scientific knowledge and understanding

1. Complete the sentences using words from the list.

| multiplied | added | difference | resultant | balanced | unbalanced |

The forces on an object can be _____ together.

The total force is called the _____ force.

If the total force is zero, all the forces are _____ .

2. Look at the diagrams.

Which diagrams show balanced forces? Choose the **two best** answers.

☐ A ☐ B ☐ C ☐ D

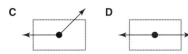

In B, the upwards arrow is half the length of the downwards arrow

3. Look at the diagram. Choose the **best** description of the total forces.

A ☐ Resultant force is zero

B ☐ Resultant force is to the right

C ☐ Resultant force is to the left

D ☐ Resultant force is upwards

4. When an object rests or slides along the ground, the weight is balanced by a force from the ground pushing up on the object called the contact force.

Look at the diagram of a heavy box being pushed along a rough floor. Label each force with the correct name from the list.

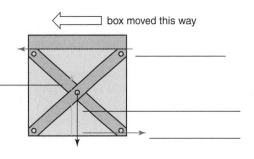

box moved this way

| contact force | weight | push | friction |

5. There are four forces acting on a cube-shaped box:

4 N to the right

3 N upwards

2 N to the left

3 N downwards

a Complete this diagram of the forces.

3 N

b Calculate the resultant force.

Total horizontal force = _____ N to the right – _____ N to the left

= _____ N to the _____

Total vertical force = _____ N upwards – _____ N downwards

= _____ N _____

So resultant force = _____ N _____ .

6. Air resistance is a force that acts when an object moves through the air.

Read each first-half sentence (a) to (d). Then match each second-half sentence to the correct first-half sentence. One line has been drawn for you.

(a) A skydiver falling through the air	**(1)** uses streamlining to reduce the air resistance.
(b) A spacecraft carrying people	**(2)** does not need to worry about air resistance, as there is no air.
(c) An aeroplane manufacturer	**(3)** uses the extra air resistance of a parachute to slow her fall.
(d) An astronaut on the Moon	**(4)** uses heat-resistant materials to protect passengers from the heat produced by air resistance on re-entering the atmosphere.

7.

The car in the diagram is travelling at a steady speed of 70 km/h.

Describe the different quantities we can measure. Use your description to state whether the forces are balanced or unbalanced. You may find some of these words helpful.

moving	speeding up	resultant	zero	balanced	speed

There is no change in the car's direction and _____.

This means the car is not _____.

This means we know that there is no _____ force acting on the car.

All the forces on the car must be _____.

8.

Practical

Challenge

Mia holds a golf ball and a larger fluffy ball at head height. She then lets go of both balls and they fall to the ground.

The speed of the balls will increase until they reach the ground.

a Explain what causes this to happen.

_____

Mia observes differences between how the golf ball and fluffy ball fall.

b Predict what you would expect to observe. Include a description of how the results should differ for the two balls.

_____

c Explain your prediction, using the ideas of balanced and unbalanced forces.

_____

_____

_____

Calculate the resultant force on the object shown in the diagram. (Remember to include the direction.)

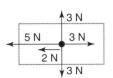

Resultant force is _____

8.2 Turning effect of a force

You will learn:

- To identify and calculate turning forces
- To plan investigations, including fair tests, while considering variables

1. Look at the diagrams. Complete each by adding labels from the words in this list.

| force | pivot | distance | load | height |

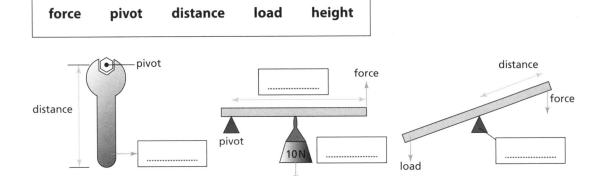

2. Complete the equations to show the relationships between **moment**, **force** and **distance from the pivot**.

a moment = _____ × _____

b $\dfrac{\rule{3cm}{0.4pt}}{\text{distance from pivot}} = \rule{4cm}{0.4pt}$

c distance from pivot = _____

3. Complete the table by writing the names and symbols of the units for force, distance and moment.

Force	Distance	Moment
newtons	_____	_____
symbol _____	symbol _____	symbol N m

4. A turning force of 5 N acts at a distance of 0.5 m from a pivot. Calculate the moment.

Worked Example

moment = force × distance from the pivot

   = 5 N × 0.5 m

   = 2.5 N m

Remember

When you calculate numerical answers you should always:

• write the equation you are going to use
• show your working
• include the units in the answer.

5. Look at these diagrams. Use the values in each diagram to calculate the moments.

(a)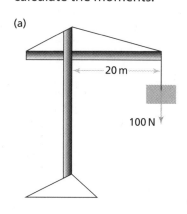

20 m

100 N

(b)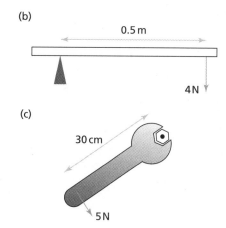

0.5 m

4N

(c)

30 cm

5N

Show Me

a moment = force × distance from the pivot

   = _____ N × _____ m

   = _____ N m

b _____

_____

_____

c _____

_____

_____

6. What is the principle of moments?

_____

_____

_____

7. Look at this diagram.

a Calculate the moment for the left-hand side of the beam.

7N   2m   3.5m   4N

_____

_____

_____

b Calculate the moment for the right-hand side of the beam.

_____

_____

_____

c Is the beam balanced? Explain your answer.

_____

_____

8. Look at these diagrams. All three beams are balanced. Calculate the missing quantity in each beam.

a _____

_____

3N   ?
3m   3m

b _____

    3 N   ?

    _____

         4 m   2 m

c _____

    4 N   6 N

    _____

              ?   2 m

9. Look at the diagram. Some devices for weighing vegetables and fruit use a balance like a see-saw. The object to be weighed is placed in the tray on the left-hand side. On the right-hand side, the different weights can be slid to different distances from the pivot.

Practical

    weights

    pivot

a Explain why the distance from the pivot to the object being weighed must be measured.

    _____

    _____

b What type of variable is the distance from the pivot to the object being weighed? Choose from the list – underline your answer.

    | independent   dependent   control |

Challenge

c You are given a bunch of bananas to weigh. You are asked to find the weight of one banana but you are not allowed to break the bunch apart. Suggest a method using the see-saw balance to find the weight of one banana.

    _____

    _____

    _____

    _____

d Yuri is given an identical see-saw balance and another bunch of bananas. His result is different from yours.

Suggest **three** possible reasons why Yuri's result is different.

_____

_____

_____

10.

Challenge

Look at the diagram of a see-saw. Ahmed wants to join in – he weighs 400 N. Explain where Ahmed should sit to balance the see-saw. In your answer you should calculate the clockwise and anticlockwise moments, and work out the difference between the two. You should then use this difference to work out how far away from the pivot Ahmed should sit to keep the clockwise and anticlockwise moments equal.

2.0 m   2.0 m

300 N   500 N

Ahmed

400 N

_____

_____

_____

_____

_____

_____

8.3 Pressure on an area

You will learn:

- To explain what causes pressure on an area
- To collect and record observations and measurements appropriately

1. Complete these equations to show the relationships between **pressure**, **force**, and **area**.

a pressure = _____

b force = _____ × _____

c _____ = _____
                                    pressure

2. Oliver investigates the pressure caused by different weights. He has a tray that measures 10 cm × 10 cm on which he places the weights. The first weight he tries is 1 N.

Worked Example

a Calculate the area of the tray in square centimetres, cm², and then in square metres, m².

Area = 10 cm × 10 cm = 100 cm²

There are 100 cm in 1 m, so there are

100 × 100 = 10 000 cm² in 1 m²

Area of tray = 100 cm² ÷ 10 000 = 0.01 m²

Show Me

b Calculate the pressure caused by the tray when a 1 N weight is placed on it.

Pressure = force ÷ area

= _____ N / _____ m²

= _____ N / m²

Remember

Prefixes tell us the size of a unit.

k = kilo-, meaning × 1000, so 1 kg = 1000 g

m = milli-, meaning ÷ 1000, so 1 mm = 1/1000 m = 0.001 m

c = centi-, meaning ÷ 100, so 1 cm = 1/100 m = 0.01 m

c Here is a table showing Oliver's results. In the first row, write in the answers to parts **a** and **b**. Then write in the area and calculate the pressure for each of the other rows.

Weight in N	Area in m²	Pressure in N/m²
1		
2		
5		
10		

_____

_____

_____

3. Blessy has a heavy parcel that weighs 300 N that she wants to place on a soft floor. The size of the parcel is shown in the diagram.

a Which side should Blessy place downwards on the floor to reduce the pressure to a minimum?

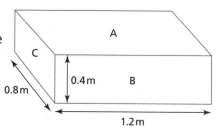

_____

b Calculate the pressure if the parcel is placed on the floor in this way.

_____

_____

_____

4. A truck weighs 120 000 N. The truck carries a load that weighs 305 000 N. The truck needs to cross an old bridge. The maximum pressure the bridge can withstand is 200 000 N/m². When the truck is fully loaded, one tyre has 0.1 m² in contact with the road surface.

Challenge

  a Work out the pressure if all the weight of truck and its load was placed on an area the size of one tyre's contact patch.

_____

_____

_____

  b Most trucks have at least two more tyres than they need to satisfy the regulations.

  Suggest and explain **one** reason why this is the case.

_____

_____

8.4 Pressure and diffusion in gases and liquids

You will learn:

- How to explain pressure in gases and liquids
- To describe the diffusion of gases and liquids
- To use results to describe the accuracy of predictions
- To describe results in terms of any trends and patterns and identify any abnormal results
- To reach conclusions studying results and explain their limitations

1. Complete the sentences below by choosing words from this list.

pressure   volume   depth   surface   particles   compressible
increases   decreases   stays the same   incompressible

Pressure in a liquid is caused by _____ in the liquid pushing against a

_____ . As you go deeper in a liquid, the pressure _____ .

Unlike a liquid, a gas is _____ . This means that if the _____ of

gas decreases, the _____ of the gas increases.

2. Look at this diagram.

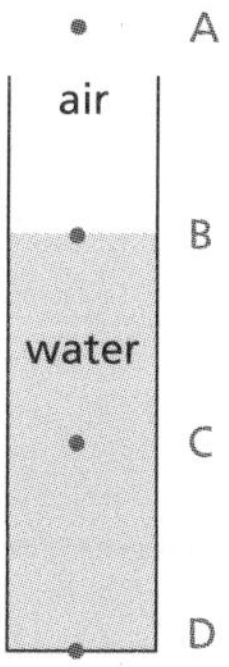

a Write down the place (A, B, C, or D) where the pressure is highest.

_____

b Describe what would happen if a small piece of light wood was placed at point B.

_____

_____

c Use your knowledge of pressure to explain your answer to part **b**.

_____

_____

d Where is the pressure lowest? Explain your answer.

_____

_____

_____

3. A piston is a round disc that moves up and down through a cylinder containing liquid or gas. The cylinder below each piston in the diagram contains the same amount of nitrogen gas. The piston can be moved up or down to different positions, as shown in the diagram.

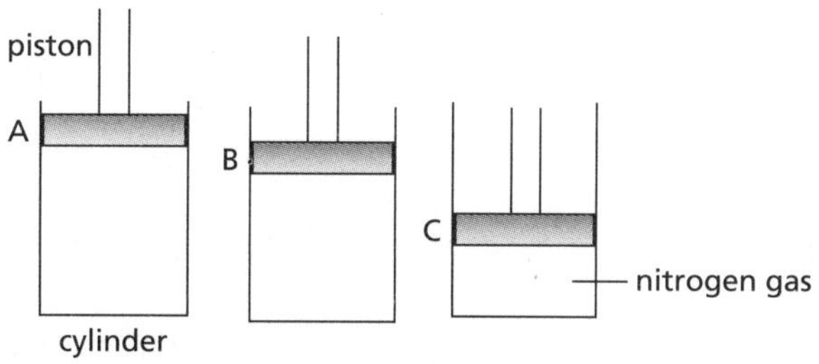

a In which position is the volume of the gas largest?

_____

b Describe as fully as possible what has happened to the volume in position C.

_____

_____

c In which position is the pressure of the gas highest?

_____

4. The pressure in a liquid can be used in hydraulic devices. The diagram shows a hydraulic device that can lift a car.

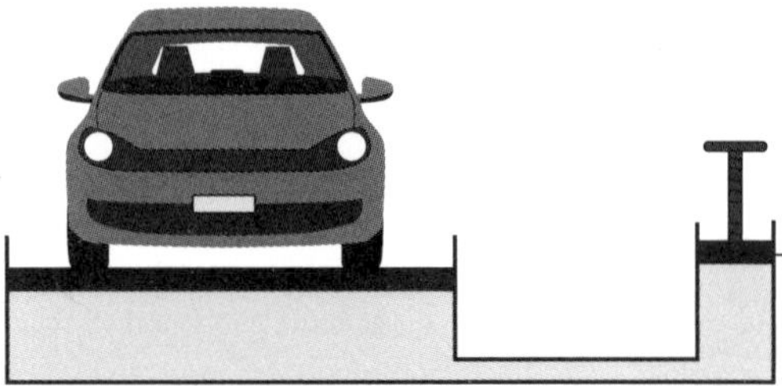

platform
area = 10A

piston
area = A

a Hydraulic devices use liquids, not gases. What property of liquids makes them suitable for use in hydraulic devices, but gases unsuitable?

_____

b Complete the following sentences by choosing words and phrases from the list. Use each word and phrase once, more than once or not at all.

| pressure   volume   force   equal to   less than   more than |

The pressure in the liquid under the car platform is _____ the pressure in the liquid under the piston.

This means that the _____ pushing up on the platform is

_____ the _____ used to push down the piston.
This makes it easier for a person to lift the car.

c Look at the different areas of the piston and the platform. If a force _F_ is used to push down on the piston, what is the size of the force pushing up on the platform?

_____

_____

5. Rajiv has a sample of gas in a leakproof container with a piston. He has a measuring device that accurately tells him the volume of the container and the pressure of the gas.

Rajiv changes the volume by moving the piston. He is very careful to make sure the temperature stays constant.

He plots a graph of his results. Use the graph to answer the questions.

Remember
Gases follow a set of relationships between different quantities. Rajiv's graph shows the relationship between pressure and volume.

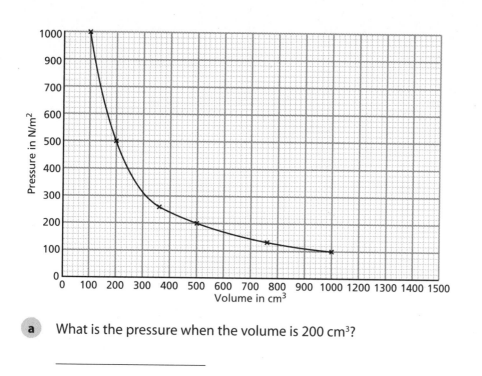

a What is the pressure when the volume is 200 cm³?

_____

b What is the pressure when the volume is 500 cm³?

_____

c Explain why it is important to keep the temperature constant.

_____

_____

d Write a conclusion that describes the pattern in Rajiv's results.

_____

_____

6. Diffusion will occur if there is a difference in:

Tick **one** box.

 Mass   ☐ Pressure

☐ Temperature   ☐ Concentration

7. The two particle diagrams show the same gas at different pressures.

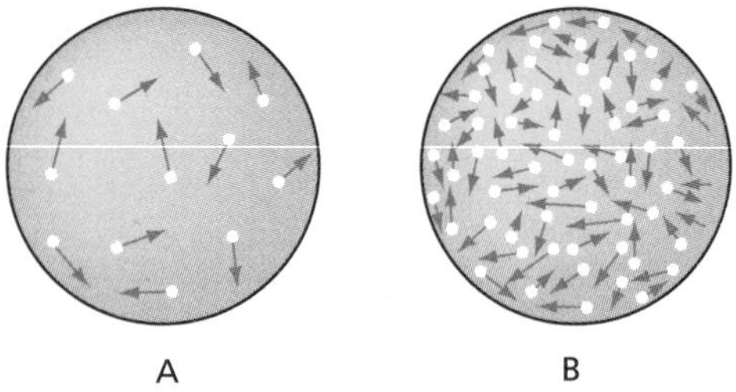

State which diagram shows the gas at high pressure.

_____

A   B

8. Which statements are true?

Tick **two** boxes.

☐ Gas pressure is a measure of particle force

☐ Gas pressure is a measure of the average particle force on the container walls

☐ Increasing the temperature causes particles to move faster

☐ Gas pressure decreases as the number of particles in the container increases

9. Complete the sentences by choosing words from the list below.

movement   slower   at the same speed   faster   diffusion   vibration

Diffusion occurs because of the _____ of particles in a gas or liquid. Gas particles move _____ than liquid particles. So _____ in liquids occurs _____ than in gases.

10. A crystal of lead nitrate was placed at one side of a Petri dish filled with water and a crystal of potassium iodide at the other. The diagram below shows an overhead view of the Petri dish.

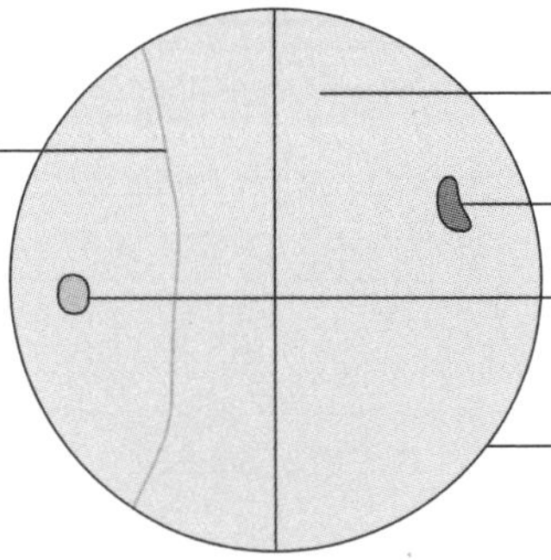

new compound ——

—— water

—— potassium iodide crystal

—— lead nitrate crystal

—— Petri dish

After a few minutes a new coloured compound is formed between them, as shown in the diagram.

Show Me

Use your scientific knowledge to explain where the new compound came from.

Both crystals dissolved _____.

The particles _____ across the Petri dish. When they met, a _____ took place and a new compound was formed.

120

11. Look at the diagram. It shows a sealed gas syringe.

gas syringe

plunger

rubber seal

a   Describe what will happen to the pressure of the gas as the plunger is pushed in.

_____

b   Use your knowledge of particle theory to explain your answer to part **a.**

_____

12. Pierre was about to go out when he noticed that his bicycle has a flat tyre.

He is looking worried because he does not know how to repair it.

Describe what has happened to the amount of air in the tyre. Suggest a reason for your answer.

_____

_____

_____

13. A perfume was sprayed at the front of the classroom.

The students predicted that the whole class would be able to smell the perfume.

A minute later the students sitting near the front could smell the perfume but the students at the back could not.

Explain this observation.

_____

_____

_____

14. Zak is getting ready for his sister's birthday party. He is blowing up balloons. His sister wants the balloons to be really big. This is causing a problem for Zak because they keep on bursting.

Explain why the balloons burst.

_____

_____

15. **Food colouring and diffusion**

Practical

Carlos and Hassan were investigating diffusion.

They put three drops of food colouring into a beaker of cold water and timed how long it took for the dye to spread out and completely intermingle with the water.

After the first experiment, Carlos made the following prediction.

As the temperature of the water increases, the time taken for the dye to diffuse through the water will also increase.

Hassan disagreed with Carlos's prediction. He thought that the time would decrease as the temperature of the water increased.

The boys collected some data by doing a second experiment and recorded it in the table below.

Temperature (°C)	Time (seconds)
20	316
30	222
40	230
50	88
60	12

Look at the results table.

a Shade in the dependent variable in red.

b Shade in the independent variable in blue.

c Draw a ring around the anomalous result.

d Write your conclusion from these results.

_____

_____

e Using your answer from part **d**, explain if Carlos's original prediction was correct.

_____

_____

122

Self-assessment

Tick the column which best describes what you know and what you are able to do.

What you should know:	I don't understand this yet	I need more practice	I understand this
The forces on an object sum to give an overall resultant force			
A resultant force of zero means all the forces on an object are balanced			
Balanced forces on an object result in steady motion in the same direction, or staying at rest			
Unbalanced forces on an object change its speed or direction of motion			
You calculate moments by multiplying the force by the distance of that force from a pivot			
You calculate moments by multiplying the force by the distance of that force from a pivot			
Moments are measured in newton metres			
If the clockwise moment and the anticlockwise moment are equal then the objects are balanced			
Pressure arises when a force is applied over an area			
To calculate pressure, use pressure = force ÷ area			
An object with a large area exerts a low pressure			
An object that exerts a high force creates a high pressure			
Pressure is caused by particles hitting surfaces			
As the depth of liquid or gas increases, so does its pressure			
Hydraulics can be used to transfer pressure from one place to another using liquids			
The particles in liquids and gases spread out from where they are at a high concentration to where they are at a low concentration. This is called diffusion			
The rate of diffusion is affected by temperature and the mass of the particles			

You should be able to:	I can't do this yet	I need more practice	I can do this by myself
Use symbols and formulae			
Describe trends and patterns in results			
Identify anomalous results			
Make predictions and review them against evidence			
Make predictions referring to previous scientific knowledge and understanding			
Make conclusions from collected data			
Compare results and evaluate methods used by others			

If you have ticked 'I don't understand this yet' or 'I can't do this yet' or mostly 'I need more practice', have another look at the relevant pages in the Student's Book. Then make sure you have completed all the questions in this Workbook chapter and the review questions in the Student's Book. If you have already completed all the questions ask your teacher for help and suggestions on how to progress.

Teacher's comments

End-of-chapter questions

1. The diagrams show forces on different objects. For each diagram, state what **direction** you would expect the resultant force to be.

(a)

40 N

(b)

20 N

20 N

(c)

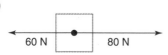

60 N   80 N

(d)

60 N   40 N

a _____

b _____

c _____

d _____

2. A worker in a factory is pushing a heavy box across a rough floor.

a Name the force that is caused by the rough floor and acts on the box.

_____

Two more workers join the first one and also push the box.

b Describe what happens to the resultant push force on the box.

_____

c Predict what you would expect to happen to the speed of the box. Justify your answer.

_____

_____

_____

d If the speed of the box increases, what happens to the force of the rough floor on the box?

_____

_____

_____

3. Moments and pressure both involve measurements of length. Use words from the list to complete the sentences and equations below.

distance   area   volume   force   mass

a **Moments**

The equation is moment = _____

b **Pressure**

The equation is pressure = _____

4. A wooden beam is placed on top of a pivot. Object A is hung from the left-hand side of the beam and object B is hung from the right-hand side of the beam. Both objects have a weight of 500 N and are placed 20 cm away from the central pivot.

20 cm

pivot

object A

object B

a Explain why the beam is balanced.

_____

b Calculate the turning force of object A.

_____

_____

c Object B is moved closer to the pivot. Predict what will happen to the position of the beam and object A, and explain your answer.

_____

_____

_____

_____

_____

5.

Practical

Hassan is investigating pressure. He puts a heavy weight on square trays of different sizes. He then puts each tray on an area of very fine sand and sees how deep the tray sinks into the sand.

a Predict whether or not Hassan should see a pattern in his results. Describe the pattern, if you think there will be one.

_____

_____

_____

Look at the table of Hassan's results.

Area of tray in cm²	Depth in mm
9	2.0
16	1.6
25	1.3
49	0.5

b Do these results fit your prediction? Explain your answer.

_____

_____

c Suggest how Hassan could improve his investigation. Explain your answer. (*Hint*: think about how easy it is to get accurate measurements using Hassan's method.)

_____

_____

d Camels live in the desert and have wide, flat feet. Based on Hassan's investigation and your knowledge of pressure, suggest why camels are well adapted to their environment.

_____

_____

_____

6. Jamila is a deep-sea diver. She needs to carry a supply of air to breathe underwater.

a Jamila needs to carry as much air as possible, but keep the container small so that it does not get in the way. Complete the following sentence using the best word from the list.

atmospheric   low   high

Jamila should carry air at _____ pressure.

b   Explain your answer to part **a.**

_____

_____

c   Explain why the air container needs very thick, strong sides.

_____

_____

7. This question is about gases.

a   When air is pumped into a tyre, the tyre inflates (gets bigger) as the air fills it.

Which statement about gases is correct?

Tick **two** boxes.

☐ The particles in a gas expand

☐ The particles in a gas have fixed positions

☐ Gases have no fixed shape

☐ The particles in a gas only vibrate

☐ Gases have no fixed volume

b   Describe what will happen to the gas pressure when more air is pumped into the tyre.

_____

c   Explain your answer to **b.**

_____

_____

9.1 Reflection

You will learn:

- To describe reflection at a plane surface
- To use the law of reflection
- To make a periscope
- To identify and control risks for practical work
- To evaluate the reliability of measurements and observations
- To describe results in terms of any trends and patterns and identify any abnormal results

1. Complete the sentences using words and phrases from the list.

> refracted   reflected   long   unchanged   lens   mirror
> straight   away from   back towards

Light always travels in _____ lines.

A light ray that falls on a shiny surface is _____.

The shiny surface is called a _____.

After it falls on the shiny surface, the light ray travels _____ the source.

2. Look at the diagram. What is the value of the angle marked X? Choose the **best** answer.

Practical

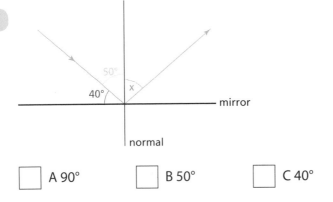

- ☐ A 90°
- ☐ B 50°
- ☐ C 40°
- ☐ D 0°

3. Write down the law of reflection.

_____

4. Youssef is trying to make a sign that people can read using a mirror. He wants the sign to say 'MOUTH' when it is reflected. Write down the order of letters Youssef should put on the sign.

_____

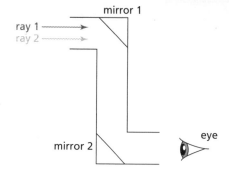

5. The diagram shows a periscope. Two light rays are shown entering the top of the periscope.

Practical

a Complete the ray diagram to show how both rays leave the periscope.

b State whether the image seen by the eye is the correct way up or upside down.

_____

6. The diagram shows a device that uses mirrors to see round corners.

Challenge

a State or calculate the values of the angles A to D.

A = _____ °   C = _____ °

B = _____ °   D = _____ °

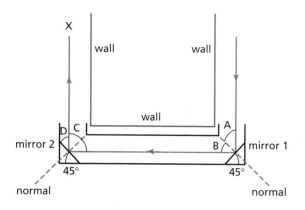

b The device is used to look at a sign that says 'HOSPITAL'. Will the image seen by the observer at point X be swapped over or still be the correct way round? Explain your answer.

_____

_____

_____

7.

Practical

Saima does an experiment to test the law of reflection in a mirror. Her results are shown in the table:

Angle of incidence (°)	Angle of reflection (1) (°)	Angle of reflection (2) (°)	Angle of reflection (3) (°)	Average angle of reflection (°)
20	21	19	20	20
30	32	28	36	32
40	41	39	40	40
50	52	49	49	50
60	60	62	58	60

a What has Siama done to check that her results are reliable?

b Draw a graph of Saima's results. Plot angle of incidence on the x-axis and average angle of reflection on the y-axis. Draw a line of best fit.

c Look at your graph. Which of Saima's results does not fit the pattern?

d Which one of Saima's original measurements was anomalous?

e What should Saima have done when she found an anomalous result?

9.2 Refraction

You will learn:

- To describe refraction of light
- To choose experimental equipment and use it correctly

1. **a** The diagram shows a ray diagram for refraction.

Complete the diagram using the words from the list.

| normal   reflection   ray   refraction   incidence   object   image |

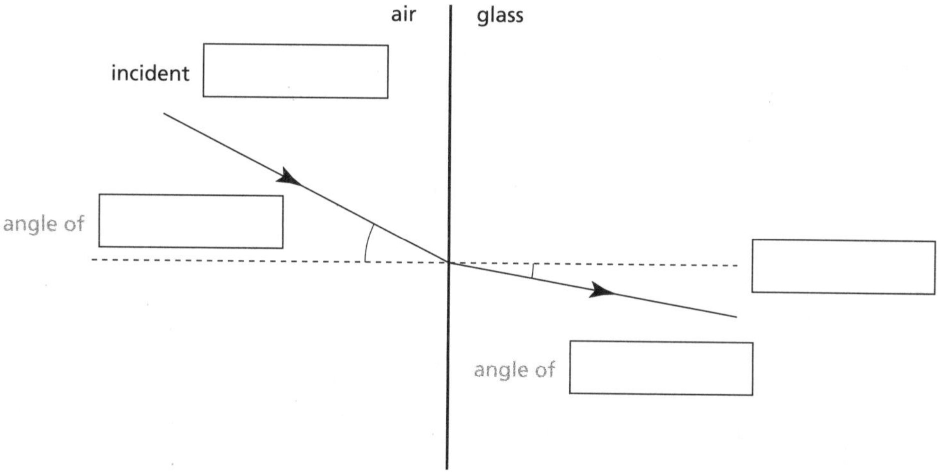

air | glass

incident [ ]

angle of [ ]

[ ]

angle of [ ]

b Why does refraction happen when the ray of light passes from the air into the glass?

_____

_____

2. The diagram shows arrangements of two different substances, A and B that are placed next to each other so they touch. The table shows the possible combinations of A and B. Choose which combinations will show refraction at the boundary between A and B. Tick [✔] which rows show refraction.

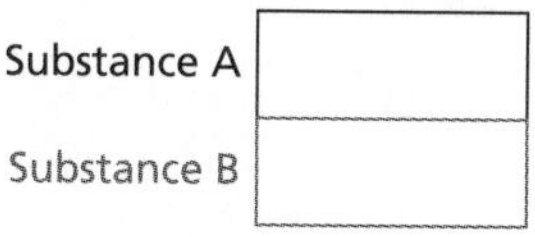

Substance A	
Substance B	

Substance A	Substance B	Does this combination produce refraction?
air	glass	☐
air	water	☐
glass	glass	☐
air	air	☐
water	glass	☐
glass	air	☐

3. In an experiment to test refraction, state the angle of refraction if the angle of incidence is 0°.

Practical

_____

4. The diagram shows the surface of a lake of water. A lamp has been placed on the bottom of the lake.

a Add a normal line at point X.

Show Me **b** Describe how the direction of the light ray changes as it enters the air.

As it enters the air, the light ray _____ .

c Complete the ray diagram to show what happens to the light as it leaves the surface of the lake. Label the angles of incidence and refraction.

5.

Practical

Safia is planning an experiment to investigate how different materials refract light by different amounts. This is Safia's diagram of the equipment she wants to use.

air

Safia needs to change the angle of incidence and measure the angle of refraction that is produced.

flashlight with
narrow beam

substance

a   Explain why it is not helpful to use an angle of incidence of 0°.

_____

_____

b   Suggest a piece of equipment Safia could use to measure the angles of incidence and refraction.

_____

c   Suggest why Safia wants to use a flashlight with a narrow beam.

_____

_____

d   What should Safia do to make sure her results are reliable?

6.

Challenge

Look at the diagram of a glass block surrounded by air. The incident ray is shown.

a   Think about what happens at point X. Draw a normal at X and the path of the refracted ray inside the glass block.

X

incident
ray

air

glass

b   Add the label 'Y' where your first refracted ray leaves the glass block.

c   Think about what happens at the point you have labelled Y. Draw a normal at Y and the path of the refracted ray through the air to the right of the glass block.

d   Add the label 'Z' to the angle of refraction between the normal at Y and the refracted ray.

e State the connection between the original angle of incidence at point X and the final angle of refraction Z at point Y.

_____

_____

9.3 Coloured light

You will learn:

- That white light is made of many colours
- To use a prism to show the dispersal of white light
- To describe what can be done with colours of light
- To reach conclusions by studying results

1. Complete the sentences using words from the list.

spectrum	absorption	reflection	refraction	scattering	dispersion

A prism can be used to separate white light into a _____.

This effect is called _____.

When light falls on a surface and no light leaves the surface, this is called

_____.

When light falls on a rough surface and the light is reflected in many different directions, this is called _____.

2. The diagram shows white light falling on a prism.

a Complete the diagram by writing in the correct labels.

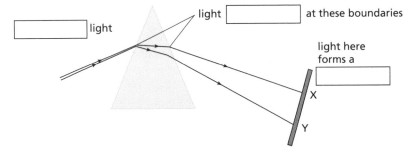

b   State the colour of the light at point X. _____

c   State the colour of the light at point Y. _____

3. The diagram shows three large coloured lamps arranged one above the other. They shine through a thick glass block onto a screen.

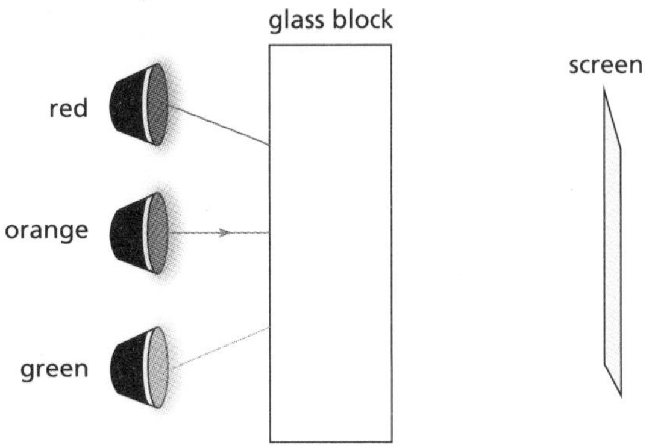

a   The orange light ray shown hits the block at a right-angle. Complete the ray diagram for the orange light. Label the point on the screen where the ray arrives with the letter 'X'.

b   Add rays for the red and green lamps to show rays that also arrive at point X.

4.

Practical

Pedro planned an experiment to test how colours of light could be mixed together. Before he carried out the experiment, he predicted which colours of light would be produced when lights were added together.

The table shows his predictions.

Light 1	Light 2	Predicted colour
Red	Green	☐ Brown
Blue	Green	☐ Cyan
Red	Cyan	☐ White
Yellow	Blue	☐ Green

a   Use your knowledge of coloured light to decide which of Pedro's predictions are correct. Tick [✔] each correct prediction in the table.

b   Pedro did not know how to produce magenta light. Predict which two coloured lights could be added to produce magenta.

_____

_____

5.

a Give the colour of a yellow shirt viewed under magenta light.

_____

b Complete the sentences below using the words in the box. Some words may be used more than once.

blue	green	secondary	adding	subtracted

Cyan light is a _____ colour and is made up by _____

blue light and green light together.

When a yellow shirt is viewed under cyan light, the shirt would appear _____.

The _____ light is absorbed by the shirt and _____ from the

reflected light, and only _____ light is reflected.

6. Look at the diagram showing different surfaces. Predict what will be observed when light falls on each surface. Complete the table with your answers.

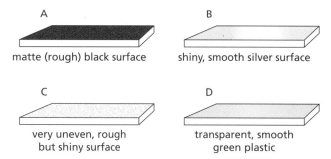

A
matte (rough) black surface

B
shiny, smooth silver surface

C
very uneven, rough but shiny surface

D
transparent, smooth green plastic

Surface	What will be observed when light falls on the surface
A	_____
B	_____
C	_____
D	_____

7. Remember what you learned about energy transfers in Stage 7 Chapter 6.

Challenge Priya designed an experiment to investigate what light does when it falls on different surfaces. She used a special red lamp to shine light on different surfaces.

Practical Priya wrote down these results in a table.

Surface	Reflected light seen	Any other observations
Metal painted matte black	None (black)	After a few minutes the metal felt warm to the touch.
Polished metal (shiny silver)	Red light	After a few minutes there was no change in temperature of the metal.
Metal polished with a green plastic layer on top	None (black)	After a few minutes the metal temperature did not appear to change.

Explain these results.

_____

_____

_____

_____

_____

_____

Self-assessment

Tick the column which best describes what you know and what you are able to do.

What you should know:	I don't understand this yet	I need more practice	I understand this
A light source gives out light			
We see things because light travels from light sources to our eyes or from light sources to objects and then to our eyes			

Light travels in straight lines			
We draw the path of light with straight lines called light rays			
A shadow forms when an opaque object blocks light			
The angle of incidence = the angle of reflection			
A periscope uses two mirrors to reflect light			
Light refracts when it travels from air into glass, clear plastic or water			
Refraction makes water look shallower from above than it really is			
A prism can split white light into a spectrum. This is dispersion			
The primary colours of light are red, blue and green			
Combining primary colours can make secondary colours and white			
Filters only allow certain colours of light to pass through them			
Filters absorb other colours of light			
Scattering takes place when light falls on a rough/uneven surface or when it hits particles in the air			
You should be able to:	**I can't do this yet**	**I need more practice**	**I can do this by myself**
Select ideas and turn them into a form that can be tested			
Plan investigations to test ideas			
Plan to be able to test if results are reliable			
Use a range of equipment correctly			
Make predictions using scientific knowledge and understanding			

Present results in tables			
Present results as graphs			
Use graphs and tables to identify anomalous results			
Test predictions with reference to evidence gained			
Compare results with predictions			
Discuss explanations for results using scientific knowledge and understanding			

If you have ticked 'I don't understand this yet' or 'I can't do this yet' or mostly 'I need more practice', have another look at the relevant pages in the Student's Book. Then make sure you have completed all the questions in this Workbook chapter and the review questions in the Student's Book. If you have already completed all the questions, ask your teacher for help and suggestions on how to progress.

Teacher's comments

· ·

End-of-chapter questions

1. The diagrams show four different experiments investigating light.

Name the effect each experiment shows.

A:_____

B:_____

C:_____

D:_____

A

glass block

B

matte black surface

C

hand   screen

D

shiny silver surface

2. Reflection and refraction are two different processes. Each sentence has two choices. Choose the correct choice for each sentence. Tick **one** box for each sentence.

a In reflection, the angle of incidence ☐ **equals** / ☐ **is different from** the angle of reflection.

In refraction, the angle of incidence ☐ **equals** / ☐ **is different from** the angle of refraction.

b In reflection in a plane mirror, the image size ☐ **is the same as** / ☐ **is different from** the object size.

In refraction, the image size ☐ **is the same as** / ☐ **is different from** the object size.

c Reflected words appear ☐ **the right way round** / ☐ **back to front**.

Refracted words appear ☐ **the right way round** / ☐ **back to front**.

3. White light passes through a red filter before arriving at a plane mirror. Then the reflected light passes through a blue filter.

State which colour of light can be seen at each of the points A to D described in the table. Choose the **best** description of the colour from the list.

white	red	orange	yellow	green	blue	black (no light passes)

Point	Colour of light
A: before red filter	_____
B: after red filter but before mirror	_____
C: after reflection but before blue filter	_____
D: after blue filter	_____

4. Some devices use prisms to direct light rays to a person's eyes. The diagram shows how these prisms can be arranged.

a Complete the ray diagram for one side of the device to show how the incident ray is changed so that it arrives at the eye.

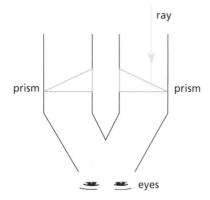

b Describe how a prism can be used to produce coloured light.

_____

_____

c State the name of this effect. Choose the **best** word from the list.

dispersion	scattering	absorption

_____

d Devices that use prisms can produce images that have coloured edges (fringes). Suggest a reason for this.

_____

_____

e One way of reducing these fringes is to place a coloured coating over the front glass of the device. State the colour of the filter you would use to **remove** red colour fringes.

_____

10.1 Magnets and magnetic materials

You will learn:

- To describe a magnetic field
- That a magnetic field surrounds a magnet
- The force exerted by a magnetic field
- The reason why Earth has a magnetic field

1. Look at the diagram of two magnets. Draw **two** arrows on the diagram to show the direction of the force acting on each magnet.

2. Look at the diagram of two magnets. Draw **two** arrows on the diagram to show the direction of the force acting on each magnet.

3. Look at the diagram of the magnetic field around a bar magnet.

a Complete the magnetic field lines.

b How does the diagram show you where the magnetic field is strongest?

_____

4. Complete the diagram below to show the magnetic field lines between two North poles. Make sure you show the direction of the field as well as its shape and strength.

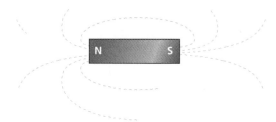

5. Complete the diagram below to show the magnetic field lines between a North pole and a South pole. Make sure you show the direction of the field as well as its shape and strength.

6. The photograph shows the magnetic field surrounding a horseshoe magnet.

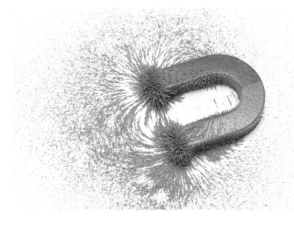

a What has been used to make the field lines shown?

_____

b What does this photograph tell you about the poles of this magnet?

_____

c What does this photograph tell you about the strength of the magnetic field surrounding this magnet?

_____

d What would be different if the poles of the horseshoe magnet were further apart?

_____

10.2 Electromagnets

You will learn:

- To describe how to make an electromagnet
- That electromagnets have many applications
- To investigate what alters the strength of an electromagnet
- To plan investigations, including fair tests, while considering variables
- To reach conclusions studying results and explain their limitations
- To evaluate experiments and investigations, and explain any improvements suggested
- To present and interpret scientific enquiries correctly
- To describe the application of science in society, industry and research

1. What is the relationship between a current in a wire and a magnetic field?

Tick (✓) the **best** answer from the list.

☐ A current flows only if a permanent magnet is placed nearby.

☐ A stationary magnet produces an electric current.

☐ A current is stopped from flowing if a permanent magnet is placed nearby.

☐ A current produces a magnetic field of its own.

2. Look at the diagram of an electromagnet. Complete the diagram by writing in the labels.

iron _____   wire_____   _____

_____

3. Describe **two** important ways in which an electromagnet is different from a permanent magnet.

Show Me

A permanent magnet produces a magnetic field all the time.

An electromagnet produces a _____ that can be _____

The strength of the magnetic field around a permanent magnet does not

_____.

We can _____ the strength of the magnetic field around an

_____.

4. **a** Tick (✓) the **one** material which could be used to make the core of an electromagnet.

- [ ] Iron
- [ ] Zinc
- [ ] Carbon
- [ ] Copper
- [ ] Mercury

> **Remember**
> Always check your answers to make sure you have written the number of reasons a question asks you for. Do not write more reasons or fewer reasons.

b Give the reason for your choice.

_____

_____

The next **six** questions (5–10) are about an experiment to investigate the strength of an electromagnet. The diagram shows the apparatus used in the experiment.

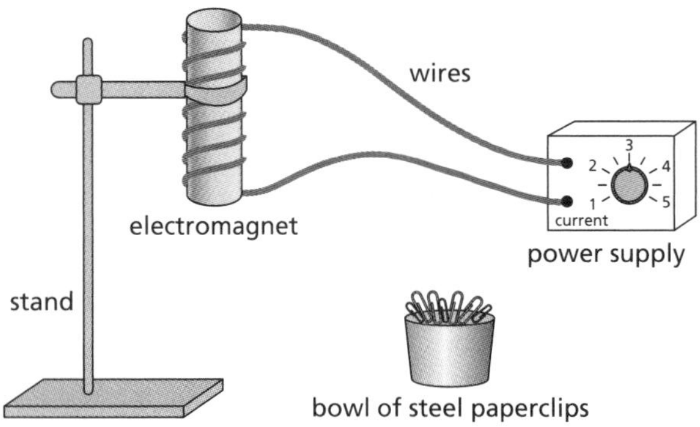

5. Jaina switches on the current and counts how many paperclips the electromagnet can pick up and hold. Jaina then repeats the experiment with different values of current.

Practical

Which type of variable is the **current** in this experiment? Tick (✓) the **best** answer.

- [ ] Independent variable
- [ ] Dependent variable
- [ ] Control variable
- [ ] Variable that does not affect the experiment

6. Jaina produces a table to record the values of the variables. She adds a heading to each column that says what type of variable each value represents. Which variable is the **dependent** variable in this experiment? Tick (✓) the **best** answer.

☐ Current

☐ Number of paperclips

☐ Material the core is made from

☐ Number of turns in the coil

7. Jaina wants to know which variables she must control in the experiment.

Tick (✓) the variables that need to be controlled.

☐ Current

☐ Number of paperclips

☐ Material the core is made from

☐ Number of turns in the coil

8. Jaina wants to show her results as a graph.

a What should she plot on the x-axis?

_____

b What should she plot on the y-axis?

_____

c How would she use her graph to check if there were any anomalous results?

_____

_____

9.

9.

Chen uses the same apparatus to measure the effect of another variable on the strength of the electromagnet. The table shows his results.

Practical

Write a short conclusion that describes Chen's results.

Current (A)	Number of paperclips lifted	Number of turns in coil	Material in core
2.0	2	6	iron
2.0	4	10	iron
2.0	6	14	iron
2.0	7	18	iron

_____

_____

_____

10.

How could Chen improve his experiment?

_____

_____

11.

Most motor vehicles include metals such as iron or steel in their bodies. When a vehicle stops working permanently or has come to the end of its useful life, it is scrapped. The vehicle is taken to a scrapyard where the different materials are separated out so they can be recycled.

Challenge

The iron and steel are squashed into a cube shape to make them easier to move around. The cube can be picked up and moved by an electromagnet.

a Discuss the advantages of using an electromagnet to move these metals around. Describe **at least three** advantages and give reasons for your choices.

_____

_____

_____

_____

_____

_____

_____

b Describe **one** disadvantage of using an electromagnet.

_____

_____

Self-assessment

Tick the column which best describes what you know and what you are able to do.

What you should know:	I don't understand this yet	I need more practice	I understand this
Magnetic field lines show the strength and direction of a magnetic field around a magnetised object			
When a current flows through a wire, it creates a magnetic field			
When a wire is coiled into a cylinder shape and a current is passed through it, it has a magnetic field like a bar magnet. This is called an electromagnet			
The strength of an electromagnet is affected by the number of coils it has, the current and the type of core used to wrap the wires around			
Electromagnets have many uses because they can be turned on or off and their strength can be changed			

You should be able to:	I can't do this yet	I need more practice	I can do this by myself
Present results in the form of tables and graphs			
Identify anomalous results			
Make conclusions by interpreting results			
Evaluate experiments and suggest improvements			
Present conclusions to others in appropriate ways			

If you have ticked 'I don't understand this yet' or 'I can't do this yet' or mostly 'I need more practice', have another look at the relevant pages in the Student's Book. Then make sure you have completed all the questions in this Workbook chapter and the review questions in the Student's Book. If you have already completed all the questions, ask your teacher for help and suggestions on how to progress.

Teacher's comments

End-of-chapter questions

1. The diagrams show magnets in four different situations. Some of the magnetic poles are labelled, but some are not.

Complete the labels of the poles on each diagram.

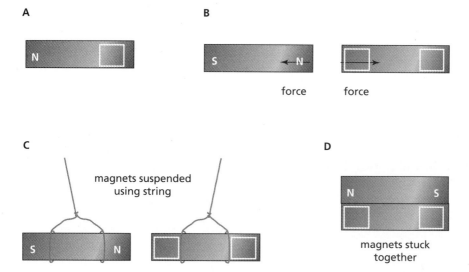

A

B

force   force

C

magnets suspended using string

D

magnets stuck together

2. The diagram shows two bar magnets placed close together. Join the dots using straight or curved lines to show the magnetic field around the two magnets. Two lines have been done for you.

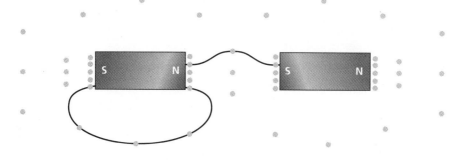

3. Drinks cans are usually made from metal. Some are made from steel. Others are made from aluminium. If we want to recycle the metals in drinks cans, we need to separate cans made from different metals.

Suggest a way of separating drinks cans. Explain the physics of your method.

_____

_____

_____

4. Lucas has had an accident in his workshop. Two boxes of nails have split and become mixed up. Lucas has a bar magnet and thinks he can use it to lift the iron nails out of the mixture. Lucas finds the magnetic field is not strong enough to pick up one nail.

Lucas has an electrical toolbox that contains the components shown in the diagram. Explain to Lucas how he could use some components to make his bar magnet stronger.

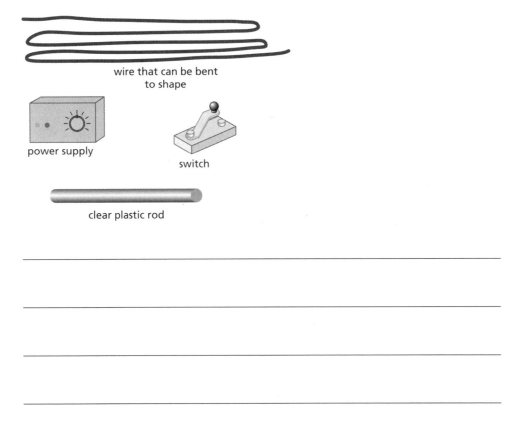

wire that can be bent to shape

power supply

switch

clear plastic rod

_____

_____

_____

_____

5. **a** The North pole of the electromagnet shown below is on the right. Complete the diagram to show the shape and direction of the electromagnet's magnetic field.

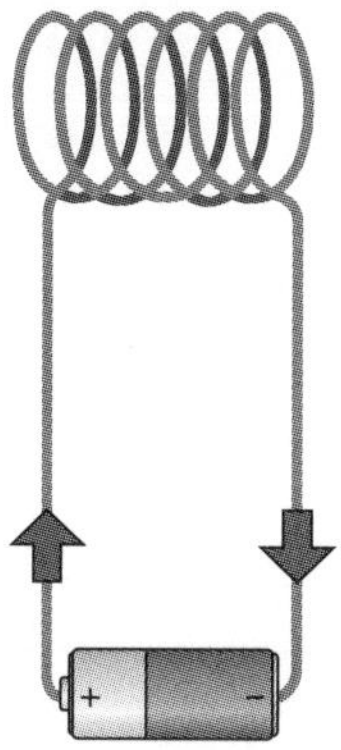

b What is the main difference between the shape of this magnetic field and that of a bar magnet?

_____

c Why do electromagnets usually have a core made from a magnetic material?

_____

_____

_____

6. Abel wants to investigate how effective different materials are as electromagnet cores. He chooses to test iron, steel, cobalt and nickel.

Challenge

a Shazia suggests that Abel should control the following variables: current, number of turns of the coil. Suggest **two** more variables that Abel could control to make sure his experiment is a fair test.

_____

_____

b Abel plans to repeat his experiment three times. Explain why.

_____

c Abel wants to present his results as a graph. Should his graph be a line graph or a bar chart? Explain your answer.

_____

_____

Earth and space

11.1 Climate and weather

You will learn:

- The difference between climate and weather
- To understand the evidence for the natural cycle of Earth's climate
- To understand how long it takes for this cycle to take place
- To describe results in terms of any trends and patterns and identify any abnormal results

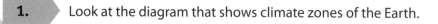

1. Look at the diagram that shows climate zones of the Earth.

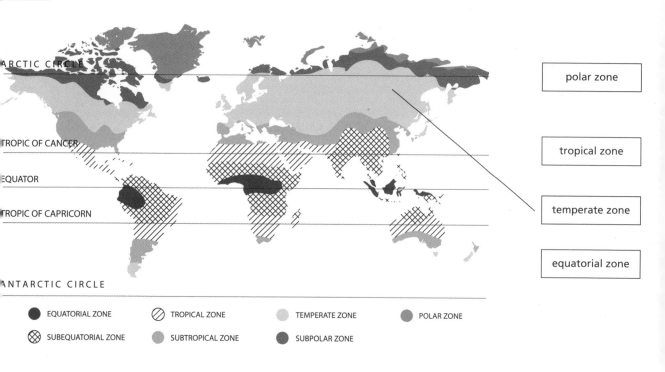

Draw lines to link each label with the correct area of Earth. One has been done for you.

2. Each sentence describes a feature of weather or climate. Choose whether a feature describes weather (W) or climate (C). Write **W** or **C** in each box. One has been done for you.

| W | A thunderstorm. |

| | **i** An increasing average temperature across Africa during the past 50 years. |

| | **ii** Heavy rainfall causing flooding in the local city. |

| | **iii** High levels of water in the atmosphere (humidity) at all times over all areas of tropical rainforest on Earth. |

| | **iv** High winds due to a tropical storm. |

3. A scientist wants to determine the warmth of the climate across the grasslands of Asia. Describe what the scientist should measure, where it should be measured and for how long. Choose the correct words or phrases to complete the sentence.

The scientist should measure **rainfall/temperature** in **one part/many parts** of Asia for **a few weeks/several years**.

4. Name the type of evidence scientists can analyse to learn about the Earth's climate tens of thousands of years ago.

A Rainfall

B Ice core

C Rock sample

D Sea level

5. Describe what happens to the amount of ice on Earth during an interglacial period.

_____

Questions **6** to **8** relate to this sketch graph, which shows the difference in average temperature on Earth for the past 140 000 years compared to today. The data for this graph came from analysis of samples taken from a place in Russia.

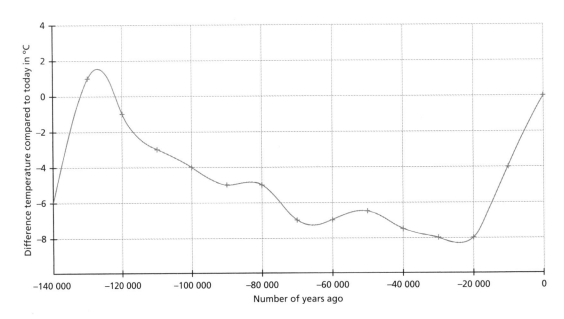

6. **a** Using the information on the graph, write down how much colder the average temperature on Earth was 20 000 years ago.

_____ °C

b Explain why the graph shows a value of zero at a time of zero years ago.

_____

7. Look at the graph again.

a What is the trend in average temperature from around 125 000 years ago to around 25 000 years ago?

A Staying about the same as today

B Decreasing

C Increasing

D Decreasing then increasing

b What is the trend in average temperature from around 25 000 years ago to today? Write one or two words.

_____

8. Look at the graph again.

a What name best describes the period of time between around 110 000 years ago and around 20 000 years ago?

A Interglacial period

B Glacial period

C Time of global warming

D Cold age

b Name the most likely cause of the changes in temperature shown on the graph.

A Natural increase in the amount of carbon dioxide in the atmosphere

B Production of carbon dioxide by human activities

C Asteroid collision with Earth

D Small changes in the Earth's orbit around the Sun

9. Describe the difference between an ice age and a glacial period.

Show Me An ice age lasts for a period of time of around

_____

A glacial period lasts for between

_____

10. Describe two types of evidence scientists can analyse to estimate when glacial periods occurred on Earth. Include an example of what each type of evidence tells us about

Challenge glacial periods.

_____

_____

11.2 Climate change

You will learn:

- To understand what can cause changes in the Earth's climate
- To identify trends in measurements of the Earth's climate
- To use scientific understanding to evaluate issues
- To discuss the global environmental impact of science
- To discuss how scientific knowledge is developed over time
- To describe how scientific progress is made through individuals and collaboration

1. Which of the following is NOT a gas found naturally in large amounts in the Earth's atmosphere?

   A Oxygen

   B Carbon dioxide

   C Hydrogen

   D Nitrogen

2. Name the gases that are described by each of these paragraphs, using the words from the box.

carbon dioxide	nitrogen	oxygen	argon

This gas makes up 78% of the Earth's atmosphere. It is not a greenhouse gas. It does not take part in combustion reactions. _____ .

This gas makes up 21% of the Earth's atmosphere. It is not a greenhouse gas. It takes part in all combustion reactions. _____ .

This gas makes up 0.03% of the Earth's atmosphere. It is a greenhouse gas. It is produced by many combustion reactions. _____ .

3. Which gas has the greatest direct effect on the average temperature at the surface of Earth?

A Nitrogen

B Oxygen

C Argon

D Carbon dioxide

4. Fossil fuels are all composed mostly of hydrocarbons. Which of the following fuels humans use is NOT a hydrocarbon?

A Methane gas

B Coal

C Oil

D Hydrogen

5. Name **three** gases that are pollutants produced by the combustion of fossil fuels.

1 _____

2 _____

3 _____

6. Explain how climate change is causing sea levels to rise.

Show Me

The average temperature around the world is _____ .

This affects the polar ice caps by _____ .

As more ice is affected, the amount of water in the oceans

_____ .

7. Methane is a hydrocarbon and a greenhouse gas.

a Explain what 'hydrocarbon' means.

 Show Me

A hydrocarbon is a substance containing only

_____ .

b Name **one** other example of a greenhouse gas. _____

c Methane taken from under the ground is used as a fossil fuel. It is burnt in air to release energy. Name the **two** substances produced in this reaction.

1 _____   2 _____

Questions **8** and **9** refer to the graph of changing amounts of a gas in the Earth's atmosphere.

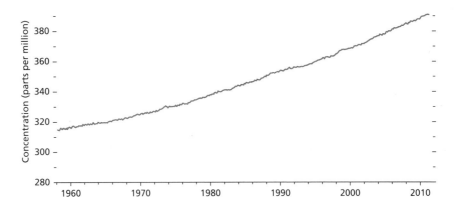

8. **a** Name the gas that is measured in the graph. _____

b Describe the trend shown in the graph. _____

c Predict the most likely trend in the graph between 2010 and 2030. Justify your answer.

_____

_____

9. **a** The graph shows changing amounts of a greenhouse gas. Describe the effect on the Earth's average temperature that this change is having.

_____

b Over what timescale have scientists predicted that the Earth's average temperature will increase by between 1.5 and 4.0 °C?

    A Under 10 years

    B Under 100 years

    C Under 1000 years

    D Under 1000 000 years

10. The following phrases describe extreme events that take place around the Earth every year. Tick a box to show which events are increasing in number and strength due to climate change, and which events are not affected by climate change.

Events	Not affected by climate change	Increasing due to climate change
Tropical storms	☐	☐
Earthquakes	☐	☐
Floods	☐	☐
Droughts	☐	☐

11. **a** Explain why measuring climate change and its effects has to involve as many countries as possible.

Challenge

_____

_____

b Suggest why it is possible for some places to have occasional colder winters than usual even though the global average temperature is increasing.

_____

_____

11.3 Renewable and non-renewable resources

You will learn:
- To identify renewable and non-renewable resources
- To describe how humans use renewable and non renewable resources
- To sort and group materials using secondary information
- To describe the application of science in society, industry and research
- To use scientific understanding to evaluate issues
- To discuss the global environmental impact of science

1. Complete the sentences. Use the words in the box.

renewable	non-renewable

Resources that cannot be replaced easily over short periods of time are _____.

Resources that can be replaced easily over short periods of time are _____.

2. Tick boxes to show which of the following resources are renewable and which are non-renewable. The first has been done for you.

Resource	Renewable	Non-renewable
Coal	☐	✔
Natural gas	☐	☐
Wind	☐	☐
Rare metals	☐	☐
Rock	☐	☐
Solar energy	☐	☐
Nuclear fuel	☐	☐

3. Number each statement so that when they are read in order from 1 to 5, they describe the formation and use of coal. The first one has been done for you.

| 1 | Organisms that lived in swampy areas died. |

| | Over millions of years, more layers of material built up on top of the remains. |

| | Humans dug mines to extract the coal from underground so it could be used as a fuel. |

| | The weight of the layers of material squeezed the remains to form hard, black coal. |

| | The remains were covered with mud and sand. |

4. Choose **two** benefits of fossil fuels from the following list. Tick **two** boxes.

| | They produce very little pollution. |

| | They are energy efficient. |

| | They are relatively cheap to extract. |

| | Their use does not increase the levels of greenhouse gases in the atmosphere. |

| | They are easy to replace. |

5. In an experiment to measure the levels of pollution from soot, a white tissue is hung outdoors in a position sheltered from rain.

Practical

a Explain why the tissue must be sheltered from rain.

_____

b Explain why the tissue should be white, not coloured.

_____

c Why does this experiment not measure all the pollutants produced by burning fossil fuels?

_____

_____

6. Draw lines to match the key scientific word with its explanation.

fossil fuels	renewable fuels made from plants
pollution	non-renewable fuels made from dead plants and animals
biofuels	the introduction of harmful substances into the environment

7. Humans have used wood from many different species of trees to construct buildings, ships and smaller items such as furniture.

a State whether wood is a renewable or non-renewable resource.

_____

b Some species of trees grow much more quickly than others. Suggest **two** reasons why it is better for people to use wood from faster-growing trees.

_____

_____

8. Iceland is an island in the north Atlantic ocean. It is positioned over the boundary between two tectonic plates. Look at the pie chart, which shows the energy resources Iceland used to generate electricity in 2015.

a What percentage of Iceland's electricity was generated using renewable resources? _____

b Calculate the percentage of Iceland's electricity that was generated using geothermal energy.

_____

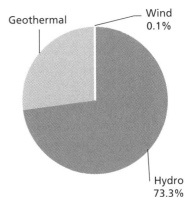

Geothermal

Wind 0.1%

Hydro 73.3%

c   Explain why Iceland can generate large amounts of electricity using geothermal resources, but many other small countries cannot.

_____

_____

9.   Most people use plastic materials every day.

a   Describe the difference between 'normal' plastics and bioplastics.

_____

_____

b   Describe how people's use of 'normal' plastics is causing problems in the environment.

_____

_____

c   Explain how bioplastics can reduce these problems.

_____

_____

10.   Scientists and engineers from around the world are investigating how we can replace the use of non-renewable resources with renewable resources.

Challenge

a   Explain how this process of change can help to limit the effects of climate change.

_____

_____

_____

_____

b Explain how this process of change should also help to prevent some resources from becoming unavailable.

_____

_____

_____

_____

Self-assessment

Tick the column which best describes what you know and what you are able to do.

What you should know:	I don't understand this yet	I need more practice	I understand this
Weather describes the conditions in the atmosphere over a small area at a particular time			
Climate describes the conditions in the atmosphere over a large area, averaged over a long time			
Earth's climate changes naturally over time, involving warm periods and ice ages			
The natural cycle of changes in Earth's climate takes place over very long time periods, typically tens of thousands of years (for glacial and interglacial periods) and hundreds of millions of years (for ice ages)			
The mixture of gases in the Earth's atmosphere can change over time			
Changes in the amounts of some gases in the Earth's atmosphere, particularly carbon dioxide, cause changes in the Earth's climate			
The Earth's climate is changing more rapidly than is natural because of human activities			
These changes include rising average temperatures, melting polar ice, rising sea levels and more extreme weather events			

	I can't do this yet	I need more practice	I can do this by myself
As humans increase their population and develop new technology, they use more and more resources for energy and for making things			
Renewable resources are resources that can be replaced easily in a short space of time			
Non-renewable resources cannot be replaced easily and will run out one day			
The use of many non-renewable resources, such as fossil fuels, causes pollution			
The combustion of fossil fuels produces carbon dioxide gas, which causes climate change			
We need to develop more renewable resources to replace the non-renewable resources we use, to reduce pollution and to find resources that will not run out			
You should be able to:	**I can't do this yet**	**I need more practice**	**I can do this by myself**
Describe the evidence for the natural changes in Earth's climate, including effects on the landscape, ice cores and fossils			
Identify trends in measurements of the Earth's climate			
Interpret observations and measurements of the Earth's climate			
Identify whether the main hypothesis about recent climate change is testable			
Plan an investigation to help gather evidence about climate change			
Sort and group ideas by using information			
Make and test predictions about investigations into air pollutants			
Make conclusions by interpreting results			

If you have ticked 'I don't understand this yet' or 'I can't do this yet' or mostly 'I need more practice', have another look at the relevant pages in the Student's Book. Then make sure you have completed all the questions in this Workbook chapter and the review questions in the Student's Book. If you have already completed all the questions ask your teacher for help and suggestions on how to progress.

Teacher's comments

End-of-chapter questions

1. Choose the correct explanation of the difference between climate and weather. Tick **one** box.

☐ **A** Weather describes an average of conditions over time and over a large area; climate describes the conditions at one moment in time in one place.

☐ **B** Weather describes the conditions at one moment in time over a large area; climate describes an average of conditions over time in one place.

☐ **C** Weather describes an average of conditions over time in one place; climate describes the conditions at one moment in time over a large area.

☐ **D** Weather describes the conditions at one moment in time in one place; climate describes an average of conditions over time and over a large area.

2. The average temperature of the Earth's atmosphere near the ground has been increasing over the past 150 years.

a What have scientists determined to be the most likely reason for this increase?

A The impact of a large asteroid.

B Humans' use of nuclear fuel.

C Humans' use of fossil fuels.

D Changes in the Earth's orbit around the Sun.

b Describe **three** effects of this increase in temperature.

_____

_____

_____

c Explain how replacing the use of non-renewable resources with renewable resources will help to limit the effects you described in part b.

_____

_____

_____

_____

3. Humans extract rare metals such as neodymium from the Earth's crust. It is expensive to find and extract these metals.

a Are these rare metals renewable or non-renewable resources? Explain your answer.

_____

b Suggest why the use of these rare metals has increased significantly over the past 20 years.

_____

_____

c Describe **two** benefits of recycling products made using these resources.

_____

_____

4. The use of fossil fuels produces pollution.

a List **three** substances that are pollutants produced by the burning of fuels.

_____

b Describe the sources of the fuels that produce these pollutants.

_____

_____

c Fossil fuels are **non-renewable**. Explain what this means.

_____

d Wind power is a renewable resource. List **three** other sources of energy that are renewable.

_____

e Describe **two** disadvantages of using wind power.

_____

_____

5.

Practical

A team of scientists working on an island in an ocean measured the height of high tide every day for 100 years. Each year, they added up all the values and calculated an average.

Some of their results are shown in the table.

Year	Average height of high tide in metres	Change over 20 years in metres
1910	3.51	
1930	3.56	0.05
1950	3.61	0.05
1970	3.67	0.06
1990	3.75	___(i)___
2010	3.85	___(ii)___

a The change over each 20 years is calculated by the equation:

Change = (average height at end of 20 years) – (average height at start of 20 years)

So change between 1910 and 1930 = 3.57 – 3.51 = 0.06 m.

Calculate the change in height:

i between 1970 and 1990: Change = _____ – _____ = _____ m

ii between 1990 and 2010: Change = _____ – _____ = _____ m

b Choose the most precise description of the trend in these results from the following sentences.

   A Average height of high tide has remained the same for 100 years.

   B Average height of high tide has increased steadily over 100 years.

   C Average height of high tide has increased by different, random amounts over 100 years.

   D Average height of high tide has increased over 100 years, and the change in height has increased every 20 years.

c Scientists using a computer model that predicts the effect of climate change suggest that by 2030, the average height of high tide will be 3.97 m.

Do you agree with this prediction? Explain your answer.

_____

_____

_____

6.

Challenge

Humans use plastic materials made from oil every day.

Evaluate the advantages and disadvantages of using these plastic materials.

_____

_____

_____

_____

_____

_____

12.1 Asteroids

You will learn:

- To describe what asteroids are
- To describe how asteroids form
- To describe how scientific hypotheses can be supported by evidence
- To describe the application of science in research
- To discuss how scientific knowledge is developed over time

1. Choose the best description of an asteroid. Tick **one** box.

   ☐ **A** Object made of rock or gas that orbits a planet.

   ☐ **B** Object made of rock or gas that is too small to be a planet and orbits the Sun.

   ☐ **C** Object made of rock that orbits a planet.

   ☐ **D** Object made of rock that is too small to be a planet and orbits the Sun.

2. Each sentence describes a feature of the planet Jupiter or an asteroid. Choose whether a feature is found on an asteroid (A) or on Jupiter (J). Write **A** or **J** in each box.

   ☐ An atmosphere.

   ☐ A surface of rock.

   ☐ A diameter larger than that of Earth.

   ☐ A diameter smaller than that of Earth.

3. Sort the following objects into order of distance from the Sun, with 1 being the closest and 5 being the furthest away. Write one number in each box.

   ☐ **A** Jupiter

   ☐ **B** Mars

   ☐ **C** Main asteroid belt

   ☐ **D** Earth

   ☐ **E** Neptune

4. Choose **two** reasons from the following list that help to explain why asteroids are more difficult to observe than planets. Tick **two** boxes.

☐ **A** They are all smaller than planets.

☐ **B** They only have thin atmospheres.

☐ **C** They produce their own light.

☐ **D** They have darker surfaces than planets.

☐ **E** Their orbits cannot be predicted.

5. The pictures show four objects found in the Solar System. Draw lines to match each object to its description.

A   **B**   **C**   **D**

rocky planet   Sun   asteroid   gas planet

6. Describe how parts of some asteroids can be found on Earth.

Show Me

Sometimes an asteroid can be pulled towards the Earth by _____.

If the asteroid enters the Earth's atmosphere, it heats up because

_____ .

Most of the asteroid _____ .

A small part of the asteroid may reach _____ .

7. The Solar System is thought to have formed about 4.6 billion years ago. When are the asteroids in the Solar System thought to have formed?

☐ **A** Over 5 billion years ago

☐ **B** About 4.6 billion years ago

☐ **C** About 1 billion years ago

☐ **D** About 460 million years ago

8. One hypothesis suggests that many asteroids in the asteroid belt were produced when a planet that was forming was pulled apart.

a Suggest how this hypothesis could be tested.

_____

_____

b Name the type of force that could cause a planet to be pulled apart.

_____

c Explain why the asteroids formed in this way would continue to orbit the Sun.

_____

_____

9. Describe **two** ways in which evidence has been found to support hypotheses about how and when the asteroids formed.

_____

_____

10. **a** Explain why it might be useful in the future to send people or robots to set up mines on an asteroid.

Challenge

_____

_____

b Describe **three** problems scientists and engineers will have to solve before we can mine materials from an asteroid.

_____

_____

_____

12.2 Stars and galaxies

You will learn:

- To describe our galaxy, the Milky Way
- To understand how the stars, dust and gas in the Milky Way orbit the centre of the galaxy
- To explain an analogy and how to use it as a model
- To use analogies
- To describe how scientific progress is made through individuals and collaboration

1. What type of object do all the stars in our galaxy orbit around?

   A Planet

   B Star

   C Asteroid

   D Black hole

2. Which of the following is the closest estimate for the number of stars in our galaxy?

   A 250 000 000

   B 1000 000 000

   C 250 000 000 000

   D 1000 000 000 000

3. Sort the following measurements into order of size, with 1 being the smallest and 5 being the largest. Write one number in each box.

[ ] **A** Centimetre

[ ] **B** Light-year

[ ] **C** Astronomical unit

[ ] **D** Metre

[ ] **E** Kilometre

4. Describe how most people can easily observe our galaxy, the Milky Way.

_____

_____

5. What shape is the Milky Way thought to be?

A A sphere (football shaped)

B A flat spiral

C A straight cylinder

D A cube

6. Explain how a force or forces cause the stars in a galaxy to orbit its centre.

Show Me

The force that causes the stars in a galaxy to orbit the centre is called

_____ .

This force acts to attract all objects that have _____ .

At the centre of a galaxy is _____ .

This pulls all the stars _____ .

_____

7. Which **two** elements are the main parts of interstellar gas? Tick **two** boxes.

[ ] **A** Helium   [ ] **B** Chlorine

[ ] **C** Carbon   [ ] **D** Hydrogen

[ ] **E** Oxygen

8.

a Describe what is meant by the word 'exoplanet'.

_____

b Explain why it is not possible to observe exoplanets using telescopes that detect light.

_____

c Describe briefly how scientists can observe the effects of exoplanets on the stars that they orbit.

_____

_____

9.

a Describe **two differences** between stars and asteroids.

_____

_____

b Describe **two similarities** between stars and asteroids.

_____

_____

10. Scientists construct computer models of how stars move in galaxies. These models use equations to calculate the different forces acting on each star and predict how they will move.

a Explain why scientists cannot show the movements of stars using normal experiments.

_____

_____

_____

b Explain why the equations used to calculate how stars move are similar to those used in models of the Solar System.

_____

_____

_____

11. The most widely accepted model for how our Solar System formed is also believed to be an analogy for the formation of a galaxy.

Challenge

a Explain what an 'analogy' is.

_____

_____

b Describe how our Solar System is thought to have formed.

_____

_____

c Explain what features of the model for the formation of our Solar System are the same as the model for the formation of a galaxy.

_____

_____

_____

Self-assessment

Tick the column which best describes what you know and what you are able to do.

What you should know:	I don't understand this yet	I need more practice	I understand this
Asteroids are rocky objects that formed at the same time as the Solar System formed			
Asteroids formed when rocks and dust clumped together because of the effects of gravity			
Asteroids are smaller than planets and do not have an atmosphere			
Our galaxy, the Milky Way, contains stars, solar systems, interstellar gas and interstellar dust			
Our Sun is one of hundreds of billions of stars that orbit the centre of the Milky Way			
The stars in a galaxy orbit a central black hole, because of the effects of the force of gravity			

You should be able to:	I can't do this yet	I need more practice	I can do this by myself
Describe evidence for the formation of asteroids			
Research and present information about asteroids appropriately			
Describe how the Milky Way can be modelled using observations of other galaxies			
Explain what an analogy is			
Use an analogy to explain how astronomers can estimate the number of stars in the Milky Way			

If you have ticked 'I don't understand this yet' or 'I can't do this yet' or mostly 'I need more practice', have another look at the relevant pages in the Student's Book. Then make sure you have completed all the questions in this Workbook chapter and the review questions in the Student's Book. If you have already completed all the questions ask your teacher for help and suggestions on how to progress.

Teacher's comments

End-of-chapter questions

1. Choose the correct definition of the unit called the light-year. Tick **one** box.

- [ ] **A** The average distance between the Earth and the Sun.
- [ ] **B** The time it takes for light to cross from one side of our galaxy to the other.
- [ ] **C** The distance light travels in empty space in one Earth year.
- [ ] **D** The time it takes for the Sun to orbit the centre of our galaxy.

2. This question is about analogies, which can be used to help explain a problem or describe an idea.

a What is an analogy? Tick the correct definition.

- [ ] **A** A comparison between two different situations or models.
- [ ] **B** A set of instructions for carrying out an experiment.
- [ ] **C** A technical diagram of apparatus.
- [ ] **D** A measuring device that uses a moving pointer.

b Describe how an analogy is useful in estimating the number of stars in our galaxy.

_____

_____

_____

3.

a Most asteroids are found in an area called the 'asteroid belt'. Describe where this is found in the Solar System.

_____

b Give an estimate of the age of most asteroids in our Solar System.

_____

c Justify your answer to part **(b)**.

_____

_____

4. Our Sun is one of hundreds of billions of stars which make up our galaxy.

a Give the name of our galaxy. _____

b Name the two main gases making up interstellar gas.

_____

_____

_____

c Scientists think that our galaxy is spiral-shaped, even though we cannot observe the shape (because we are inside the galaxy looking outwards). Describe the evidence scientists have collected that suggests our galaxy is spiral-shaped.

_____

_____

_____

5. The table shows data about objects in our Solar System.

Object	Distance from Sun in millions of km	Distance across object in km	Shape of object	Object contains
A	0	1 390 000	Sphere (football-shaped)	Gas
B	500	25	Peanut-shaped (like two footballs stuck together)	Rock
C	150	12 700	Sphere	Rock
D	780	140 000	Sphere	Gas

a Use the data to suggest what objects **A** and **B** are.

_____

b Explain your answers for **A** and **B**.

_____

_____

Challenge   c Use the data to suggest what objects **C** and **D** are.

_____

d Explain your answers for **C** and **D**.

_____

_____